WOULD YOU PUT YOUR HEAD IN A MICROWAVE OVEN

by
Gerald Goldberg, MD

Bloomington, IN Milton Keynes, UK

authorHOUSE

AuthorHouse™
1663 Liberty Drive, Suite 200
Bloomington, IN 47403
www.authorhouse.com
Phone: 1-800-839-8640

AuthorHouse™ UK Ltd.
500 Avebury Boulevard
Central Milton Keynes, MK9 2BE
www.authorhouse.co.uk
Phone: 08001974150

This book is a work of non-fiction. Unless otherwise noted, the author and the publisher make no explicit guarantees as to the accuracy of the information contained in this book and in some cases, names of people and places have been altered to protect their privacy.

First published by AuthorHouse 2/14/2006

ISBN: 1-4259-0480-7 (sc)

Printed in the United States of America
Bloomington, Indiana

This book is printed on acid-free paper.

TABLE OF CONTENTS

FORWARD

I would imagine that you would have answered NO for the obvious reason that you are aware that microwave radiation can damage your body. The average consumer who has used a microwave oven understands it potential for causing harm and is grateful that it is packaged in a way to protect them.

This book focuses on the health consequences that are arising from having created a planetary microwave environment. As an outgrowth of our current satellite communication technologies we have transformed our atmosphere into a sea of microwave radiation and are immersed in it. The health consequences of continuous microwave exposure are cumulative. This book explores the established patterns of illness that are known to be produced by microwave radiation and the signature of that pattern.

The simultaneous rise of illness across broad geographic areas reflects this pattern of immersion. Ask yourself what is the common denominator. Wittingly or unwittingly we have:

"Transformed the atmosphere

Of this planet

Into an

<u>Open microwave system"</u>

This manuscript was written to present the data to the average reader so that they could judge the issue for themselves.

The health consequences from continuous low level exposure from microwave radiation can be catastrophic. The transmission of microwave transmission can be handled in a safe and effective manner to provide for global communication without affecting the overall health of our planet.

The book starts by examining dosimetry which shows which parts of the body are most sensitive to microwave radiation based on studies done by our Armed Forces. Dosimetry is also predictive of the pattern of injury that one would expect to find.

The second chapter explains how microwave radiation produces disease. It examines the mechanisms of disease production in living systems. It also explains how microwave injury manifests as specific organ system dysfunction, be it neurological, behavioral, immune, cancer etc.

The third chapter links the rise of radiation against the simultaneous rise of certain illnesses in society. It demonstrates through the use of graphs that the parallel and symmetrical rise of certain illnesses across broad geographic areas reflects common force acting equally on all these regions. The graphs also demonstrate that it is the rate of increase not the absolute number of cases in a given region that is important. It is important to emphasize that microwave radiation is not the only cause of disease. It is potentially one of the few common links that unite large segments of our population to recreate and produce similar patterns of illness in widely separated geographic locations.

Microwave injury can also contribute to other patterns of illness. In the fourth chapter, under medical equivalents, I have grouped the common names that are attributed to illnesses by organ systems. If common illnesses are examined from the perspective of simultaneous parallel geographic increase than the influence of microwave injury can be discerned.

The fifth chapter lists common available supplements that have been shown through medical research to be of value in offsetting microwave radiation injury or damage. Including in the listing are several references for each supplement listed. Most of the items are readily available.

The use of microwave technologies increases daily. This should be obvious to anyone who is aware of the rise in global communication capabilities. The health risks and consequences are already well documented. Microwave radiation is present.

It is well established in the medical and scientific literature that there is a lag period of approximately 5 to 7 years from the time of initial exposure to radiation to the development of full blown disease.

Microwave radiation does not leave any tell tale residues to let you know that it has caused disease. What distinguishes microwave radiation is the pattern of injury that it produces, which is recognizable from the dosimetric and scientific studies, some of which are presented in this book.

The point to appreciate in this analysis is not in recognizing whether or not there is a smoking gun, which is obvious, but that we are all being shot.

Microwave radiation poses an extreme public health risk that may become fully realized in the next 5 to 7 years. At the point that the epidemic becomes fully realized up to 60 to 75% of the general population may become affected and incapacitated. We are potentially faced with a health crisis that will cripple most industrialized nations. If these estimates are correct than we are on the verge of a crisis that our current health infrastructure will be incapable of dealing with. The purpose of this book is to increase public awareness, facilitate corroborative interpretation of epidemiological data and to formulate preventive strategies before this epidemic comes to fruition.

CHAPTER 1:
DOSIMETRY STUDIES

Dosimetry studies reveal what parts of the body most easily absorb the radiation and become damaged. Dosimetry studies establish the pattern or signature of injury that would result from microwave exposure. An individual will cook in a predictable pattern. This will show up in a pattern of disease that will progress and can worsen over time if not corrected. The responsibility or skill that a good physician brings to bear is that they are trained to recognize patterns of illness.

This skill can also be afforded to the average citizen. One does not have to be specially trained to recognize a bruise or a cut or a puncture wound or any serious injury. Most individuals rely on the pattern of a disease to establish that they are hurt. A dosimetry model can be an invaluable tool to help the individual see the correlation between radiation exposure and disease.

The point is if one looks at the rise in the regional number of claims for related illness across this country and compares them to the rise in the use of microwave technology one is immediately struck by the immediate correlation between these two. There are not many things in nature that can produce a similar pattern of illness across broadly unrelated geographic areas of the country or the world for that matter.

The value of the use of the dosimetry model is that it clearly shows what type of damage to expect. It takes away the guess work and quickly establishes the pattern of illness that one would expect to find. Dosimetry alerts us to the fact that there are unseen forces contributing to our overall illness patterns. Any organization or company that works with radiation is required to monitor the level of exposure using this technique. This model has been adapted to look at levels of tissue exposure to radio frequency radiation. What I have reproduced here is a simplified visual reference that one can use to see what parts of the body are affected and at risk. The pattern of injury shown here is the signature of the causative agent; microwave radiation.

Firstly, I will discuss the dosimetry model and explain what it suggests. Included below is a copy of the dosimetry model that was established by the United States Department of Defense. This study was conducted by the Radio Frequency Radiation Branch of the US Air Force Research Laboratory at Brooks Air Force Base in Texas.

The site can be readily accessed through the internet by typing the URL: http://www.brooks.af.mil/AFRL/HED/hedr/. Note the dosimetry chart which appears at that website is in color. The chart uses the color spectrum of light to reflect the relative sensitivity of tissues. The organs which are most sensitive are red in color and the least sensitive are blue.

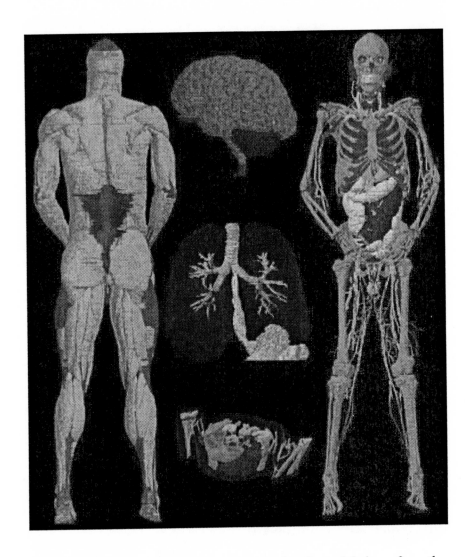

Areas of high radiation absorption/sensitivity for the most part appear darker in this representation; this also means that these areas are the most susceptible to radiation injury. These areas include the nervous system: the brain, spinal cord and major sense organs (eyes and inner ear). Also, cartilage equally absorbs high levels of microwave radiation.

Cartilage makes up the support structure for the ear lobes, nose, larynx, bronchi, respiratory tree and the connections where the ribs join the breast bone and at all the joints in the body; i.e. shoulder, elbows, wrist, finger joints, hip joint, knees, ankles and toes. Nerve damage can present as a wide range of illnesses, including cancer, including laryngeal cancers in men, which is presented in the chapter on geographic distribution.

The area of the body which has the second highest absorption level is the skeleton. The bones of the skeleton contain marrow which is the major production center for the formation of red and white blood cells. Leukemia, a cancer that predominantly affects children arises from the malformation of white blood cells from the bone marrow. This increase in radiation absorption is reflected in a rise in the incidence of these cancers.

Next in sensitivity are the teeth, lymphatics, colon, arteries, esophagus, stomach, and peripheral nerves of the body. The reader is referred to the chart showing the rise in Lymphatic cancer or lymphoma (Hodgkin's disease) in the chapter on geographic distribution.

Next in order of vulnerability are the bronchiole tissues of the lung, gonadal tissue (ovary or testes), prostate, uterus, pancreas, spleen. These are examined in the chapter on geographic distribution of cancers. Refer to the charts on the rise in the incidence of cancer of the ovaries, testes and prostrate.

Muscles and connective tissue including fascia appear to be more resilient to microwave radiation. Minor muscle damage can present as generalized myalgias (muscle pain) or weakness. Muscles are the major tissue element of the circulatory system. This includes the heart, arteries, capillaries and veins.

The most resistant organs of the body appear to be the lungs, liver, small intestine, and thick connective fascia of the lower back and the cerebellum of the brain.

The dosimetry chart is just a simplified way of showing what parts of the body would be most easily affected by continuous exposure to microwave radiation. It also helps one to spot a correlation with the rise in disease incidence with the type of illnesses one would expect to encounter. The rise in incidence of certain diseases would be uniform across broad geographic areas and would tend to follow a similar upward curve. The upward uniform sloop of the curve would reflect the effects of cumulative and continuous exposure that at the same time would parallel the rise in the utilization of this technology. For example if one sees a uniform general rise in the incidence of brain cancers, blood cancers or laryngeal cancers across broad geographical areas that exceed other types of cancers than this is an indicator that microwave radiation a contributory factor.. For further details refer to the chapter on geographic indicators, under the distribution of cancers.

Additionally the following chapter on disease promotion lists common diseases and symptoms that have been found attributable to microwave radiation.

CHAPTER 2:

DISEASE PROMOTION

1. Mechanisms of Injury

What is unique about microwave radiation is its ability to deeply penetrate into tissues and to cook or heat them up. This is very obvious to anyone who has used a microwave oven to defrost a 10 pound roast or simply to heat up a cup of coffee or pop a bag of popcorn.

Why is there a problem with slow exposure or low level exposure?

Well several things are obvious.

1. Regardless of the dosage on receives the radiation will travel deeply into the body to affect tissues.

2. Regardless of the type of exposure the effects of the radiation are cumulative. That is if you received a large exposure over a short period of time or if you received a low dose exposure over a longer period of time the results are the same. The total exposure is cumulative; in essence there is no safe dose.

The implications of these statements are several fold. So what tissues are affected and what type of damage are they prone to.

1. Fatty and solid tissues are affected the most. Note on the previous dosimetry chart the organs that are colored red. These include the most important organs in the body.

2. Heating or exposure to microwave radiation increases oxidative damage. Additionally microwave radiation tends to use up certain critical antioxidants especially catalase and SOD. Animals with decreased levels of these antioxidants have been known to have shortened life span.

3. Heating or exposure to microwave radiation decreases cellular metabolism.

4. Heating causes blood vessels to go into spasm and can shut off blood flow.

5. Heating causes burn damage deep within tissues. This is similar to a second degree burn which causes blistering. When it occurs deep with tissue it can produce a fluid filled space called a cyst.

6. Microwave radiation can disrupt cell signaling, this is noted in the electroencephalagm, EKG, and in many tests which reveal decreased electrical functioning of cells along with early fatigability.

2. Oxidative damage

Increased oxidative damage or for that matter burn damage can result in a shorter life span. Consistently in the scientific literature has been noted that animals and humans subjected to microwave radiation show decreased levels of SOD, catalase, glutathione, Coq10 along with evidence of increased byproducts of oxidative stress to cell membranes, MDA. Depletion of anti-oxidants has been shown to be an independent risk factor in the development of cancer and other illnesses associated with aging.

SOD, catalase, glutathione and Coq10 are substances that the body produces to protect itself. These substances are considered anti-oxidants. The body is continuously producing energy. Energy production produces heat and causes free electrons to be generated. The challenge that the cell had to overcome was how to cook within itself. Overcooking would cause the cell to literally cook itself. The difficulty issue for the cell is that if you burn down your own house, than you simply can not continue to cook (produce energy). In a simple manner anti-oxidants act much like insulation that one would have around a kitchen. It is desirable to have an oven or a furnace. However it is not desirable to burn down the house. This is the equivalent role of anti-oxidants. If there is excessive damage to a cell, than this will be reflected in lower anti-oxidant levels. The anti-oxidants are getting used up faster than they can be produced or replaced. The levels of anti-oxidants thus become a measure that the body or cell is overheating.

The body is continuously involved in a balancing act. The balance is the ability to generate energy and the ability to cope with burn damage. Anti-oxidants protect the structural parts of the cell from damage; these include cell membranes, structural proteins, and the genetic material of the cell. If anti-oxidants are deficient than the cell will literally age or degrade at a very rapid rate. Sustaining adequate levels of anti-oxidants is a critical part of the diet. The level of anti-oxidants in the body play a critical role in determining the amount of work an individual can do. The levels of anti-oxidants in the body have been shown to be the most sensitive indicators in stress, aging, infections, and various other disease states. They are considered the most sensitive markers of aging used by investigatory science. Refer to the references at the end of this chapter and the end of the last chapter for more complete references on this matter and the role of anti-oxidant substances in conferring protection to the cell.

Catalase is critical in reducing hydrogen peroxide to plain water. This prevents the buildup of too much H2O2 which can damage cell membranes. Notably this is critical in the perixosome, nucleus and the mitochondria. Decreased catalase levels have been found to be an independent marker of cell aging. Decreased levels of catalase in the mitochondria (cell furnace) have been found to be one of the most accurate predictive markers of accelerated aging and premature death in multiple animal models.

Microwave radiation has been shown to produce marked disruptions in energy production in cells. This has been noted as well as decreased ATP and creatinine phosphate production. ATP is considered to be the most important molecule that allows for the transfer of energy in the body. It is kind of like a miniature biological battery within the cell. If you do not have enough batteries than the cell stops working. Simple enough. The studies which are mentioned at the end of this section, repeatedly found that microwave radiation decreased the ATP content of the brains of the animals that it studied. A brain that looses the ability to generate energy (ATP) ceases to work efficiently or at all. The end result is that one can become neurologically or behaviorally impaired. This is what one finds in a child who is autistic, has ADD or other behavioral problems where the issue is impaired concentration. The same findings are also seen in individuals with psychiatric diseases, extremes of anxiety and depression as well as in dementia and Alzheimer's disease. These findings have been consistently found in all tissues examined. The scientific literature reflects that the damage can be attenuated or slowed down by the use of certain anti-oxidants (refer to the last section) and herbal products. Ultimately the body must rely on its diet to replace nutrients; it really has no other choice. Central to this point is that maintaining an adequate level of anti-oxidants in the diet, refer to the last section, will confer some degree of protection to the individual from continuous exposure to microwave radiation as well as other factors.

3. Brain function

Heating or slow cooking affects the blood circulation. Blood vessels go into spasm and produce a myriad of problems. ALS or amyotrophic lateral sclerosis is a disease which is characterized by scarring of the side of the spinal cord. Several Scandinavian studies have shown a clear link between microwave radiation exposure and an increased incidence of ALS. It has been noted that ALS appears more frequently in individuals who are exposed to microwave radiation.

The direct heating of blood vessels will result in spasm which can masquerade as various forms of headaches, either simple or migraine. Many of these headaches are attributed to stress but they may have another etiology

Other parts of the brain can also be affected. This can be due to a direct effect by microwave radiation causing spasm or blood vessels, impairing the uptake of nutrients to critical centers of the nervous system. Also microwave radiation can produce direct damage to the cells of the brain and sense organs themselves. If nerves are affected than neuropathies occur which affect movement and sensation in the body. If the brainstem is affected it can produce a host of problems, ranging from problems with balance, sleep, hearing, vision, appetite and hormonal production. If the sense organs are affected it can produce problems with vision, hearing or other sense modalities. Additionally microwave radiation has been clearly linked to DNA damage in the brain as well as an increase in brain cancers. See the references below and the section on cancer.

4. Eye problems

Visual problems can occur from an interruption of blood flow to the eye along with direct damage to the structures of the eye. Microwave radiation will cause degradation of the structural proteins and glycoproteins of the eye. Proteins and glycoprotein are critical to maintaining the structural integrity of the eye.   Microwave radiation damage to the eye can present under many guises such as macular degeneration, different forms of retinopathy, vitreous detachments, cataracts and a generalized deterioration in visual acuity. Indeed it has been well documented that direct exposure amongst chefs to prolonged microwave exposure has increased their risk of cataract formation. Also there is a growing body of evidence looking at pathology of macular degeneration and premature aging and destruction of the cones in the eye which are important in transmitting color. There is evidence that the findings of disruption of cones in the eye, a lack of stored sugar in the cones (depletion of glycogen stores in the retina), disruption in the array of the cones (to properly convey sight the cones have to be arrayed in rows and the formation of drusens (which are really fatty blisters in the inner lining of the eye) are all reproduced by microwave injury. The cones are the cells that sense color, they have to be arranged in an orderly fashion to pick up a sharp image and rely on adequate stores of sugar, glycogen to do work. The ability for a cell to store glycogen is equivalent to having a pantry in your kitchen. It signifies the storage of food so the cell can do work. Other than that the cell has to depend on the levels of sugar in the blood stream.

The pattern of injuries are very similar, though they are given different clinical terms. They all consist of a disruption of blood flow to the eye, a disruption of the architecture of the eye and changes in the light sensitivity of the eye. There is an increased incidence in drusen formation and macular degeneration, which are changes that are seen with increased oxidative damage to the eye. Additionally there is seen an increased incidence in cataracts and melanomas of the eye. The incidence of these various illnesses has gone up exponentially in the last several years. Could this be a coincidence? Across the whole nation? The increased utilization of medical services can serve as a marker of a missed epidemic. See the last chapter for possible antioxidants protecting the eye.

5. Behavioral and Cognitive Disturbances

More insidious in nature and devastating to the individual and his family is the gradual or abrupt decline in cognitive functioning and impairment in social interaction which gets written off as a genetic illness or the normal consequences of aging. Medical science seems to be incapable of addressing what constitutes normal aging in a toxic environment. Many cognitive, behavioral or neurological disease fall under the rubric of psychiatric illness with no exploration of why the overall incidence of cognitive dysfunction is increasing in modern industrialized societies. Indeed cognitive dysfunction seems to be increasing exponentially instead of in a linear fashion which is what one would expect with purely genetic disease.

What has been clearly documented in the scientific literature is that many diverse psychiatric and medical illnesses appear to share the same pattern of disrupted blood flow. The pattern is the blood flow to the left side of the brain is diminished or impaired. The left temporal-frontal region at the front of the brain seems to be most severely affected.

What is critical about this region of the brain is that it is crucial in allowing the individual to acquire and develop insights, to process information, form judgments and react to the environment. The centers involved allow for the integration of emotional learning along with intellectual development.

Studies looking at blood brain flow in individuals with ADD, autism, mania, depression, schizophrenia (both types), post traumatic stress syndrome, hyperirritability syndromes, pre-senile dementia, senile dementia, and Alzheimer's demonstrate the same exact pattern of defects. These findings show the same pattern of decreased blood flow to the parts of the brain necessary for the acquisition of insight.

These studies also imply that some of these disorders are reversible if the underlying lack of blood flow is dealt with.

It has been amply documented that microwave radiation can cause spasm in exposed blood vessels. Given the abrupt rise in psychiatric, neurological and behavioral disturbances what one may be witnessing are the effects of a national microwave lobotomy.

The left frontal region of the brain may be the most sensitive to stress from whatever the cause and has the greatest tendency to shut down. These findings reveal a common pathology which may be reversible if addressed.

Many of the current therapeutic approaches utilize cognitive retraining with the use of medication. The intent of therapy is to help the individual to develop insight. The irony of the approach is that the very part of the brain that is necessary for developing insights and integrating information are shut off!

Approaches using biofeedback to address brain wave function has been shown to be a practical alternative therapy. Instead of focusing on processing information the individual is taught to auto regulate their brain wave pattern which seems to improve the circulation problem. Improving blood flow correlates with improved information utilization. See the reference section at the end of this chapter.

Judgmental impairment resulting from impaired blood flow can also be detected by abnormal patterns in the brain EEG. The EEG shows the electrical pattern of brain activity.

In EEG studies of the brain exposed to microwave radiation a consistent finding is the loss of preparatory slow wave potential or psp waves. Psp waves are commonly absent in infants and old adults. Psp waves occur before initiating a response. The presence of psp waves seems to indicate that the person is recruiting, organizing and drawing on other areas of experience in their memory banks before initiating a response. This evidence of the absence of recruitment seems to be lost in judgmental impaired individuals. They have lost the ability to respond in a complex and meaningful way to any stimuli. It is if you took someone with a lifetime of experience and wiped their slate clean. Also microwave radiation diminishes the ability to lay down memory tracts and interferes with the metabolism of the brain.

In spite of ones good intentions in trying to reach someone who is judgmentally impaired the difficulty for any family member is that they are dealing with an individual who lacks the ability to develop insight into their behavior.

A similar loss of electrical activity has been noted in the retina of the eye and in the ear of animals that have been exposed to microwave radiation. This findings show how microwave radiation contributes to the diminished ability to organize and make sense out of incoming sensory information. The end result is that the information becomes distorted and the individual has a hard time processing the information if at all. Is it any wonder that someone so exposed would be learning disabled? There are concerned parents who are currently resisting the attempt to place microwave antennas in or around schools for these same reasons.

This chapter attempts to show how microwave radiation can exacerbate and contribute to intellectual decline regardless of the age of the individual. The last chapter in this book deals with some preventative strategies that can be useful in offsetting some of these issues. Attempting to correct or ameliorate contributory influences producing a shut down of blood flow to the brain would go a long way in correcting this problem.

6. Neurotransmitter Dysfunction

Microwave radiation can affect the chemistry of the brain directly. It acts through several different mechanisms. The first is direct DNA damage. The second is through direct disruption of metabolism and the shut down of the production of ATP and creatine phosphate. The third category is direct oxidative damage. The fourth is disruption of neurotransmitters in the cholinergic system of the brain, which includes GABA, acetylcholine, and diazepam like receptors which all tend to produce a calming effect. The fifth is a disruption of protein metabolism which is necessary to help lay down memory tracts.

Thus the affected individual has a short circuited brain which is not working well and is judgmental impaired, incapable of insight and self repair and lacks the chemical ability and sensitivity to calm itself down. Sound familiar?

Studies have repeatedly shown that the metabolism of the frontal lobes and the hippocampus are shut down. Their ability to use or produce calming neuro-transmitters are impaired. These areas of the brain are crucial in developing insight and evolving memories. Medications may ameliorate this condition but they may not be dealing with the underlying cause which is epidemic in proportion.

It has been shown that the most reliable markers of stress related injuries affecting the nervous system are lowered levels of anti-oxidants. These studies also suggest that replacement therapies are the most effective in addressing this metabolic imbalance, apart from removing any noxious influences.

7. Disruption of Blood Flow

Blood vessel spasm can impair function in any organ system in the body. The profound changes that one sees in functioning are really the end result of decreased circulation to that organ. The most familiar and dreaded example of this is blockage is to the blood vessels of the heart.

Any organ system in the body can be affected which includes the thyroid, liver, thymus, liver, spleen, pancreas, kidneys, muscles, joints, bone, lymphatic tissue or reproductive tissues. These organs are critical to the proper functioning of the body. Blood vessels of various sizes have been shown to be affected by microwave radiation. The affects can be intermittent spasm which would not show up on most commonly used radiological studies which include CAT scans and MRI. Investigational studies utilize PET scans and SPECT scans. These scans have been found to be the most reliable at looking at alterations in metabolism and blood flow. At the current time these technologies are not generally available to the public. I have seen anecdotal evidence that an individual destroyed their kidney function after two days of direct exposure to an open microwave oven. Apart from the bulk of the studies which have examined the effects of microwave radiation on the nervous system, there is a large body of literature showing that the kidneys are also greatly at risk, though this is not generally appreciated. Refer to the references at the end of this chapter and in the last chapter under melatonin and gingko bilboa.

8. Energy metabolism/Endocrine dysfunction

Microwave exposure tends to have a direct effect on energy metabolism in the body. As the body heats up, due to the direct effect of microwave radiation, it attempts to compensate by down-regulating or shutting off its own energy production. In response to increased oxidative damage, that is cooking the cells, the cell will upgrade its anti-oxidant defenses while at the same time shutting down energy production.

This occurs through several mechanisms. ATP production is shut down. Anti-oxidant enzymes are shunted from the mitochondria to participate in oxidative defense usually at cellular membranes. These findings have been reproduced in many studies looking at the effects of microwave radiation damage to the brain. ATP and creatine phosphate levels were noted to be diminished after microwave radiation exposure. These are two of the most important energy producing molecules in the brain. ATP is considered central in providing any cell with the ability to do work. Literally ones brain begins to fry. The brain begins to shut down. An individual may notice a sense of brain fatigue, loss of alertness, impaired memory or concentration. This has been clearly shown in animal models. Imagine what the effect would be in a young child, or in someone with some slight degree of mental impairment, say an elderly individual. That individual would become refractory to any intervention and would literally appear to be out of their mind.

The net result of microwave exposure is that the body heats up, while energy metabolism shuts down. Several adaptive mechanisms get triggered at the same time. The thyroid gland attempts to compensate by shutting off. The mitochondria, which are the furnaces of the cell, start to shut off. The respiratory enzymes which constitute the fuel source of the mitochondria are shunted elsewhere in the cell to bolster the cells anti-oxidant defenses. Some of the respiratory enzymes are anti-oxidants and have a dual use in the cell. The cell will wisely, if under oxidative attack, shut off it mitochondria to limit burn damage, while at the same time freeing up its antioxidants to limit oxidative damage. This clever solution insures that the cell will maximize its attempt to maintain its membrane integrity and survivability while at the same time limiting damage from within. At the same time cortisol levels will rise in the body. Cortisol levels rise with stress and have powerful anti-inflammatory effect.

The end result is that the individual loses their ability to regulate their own temperature. They become extremely susceptible to extremes of temperature, either hot or cold. One metabolically starts to behave as though they were reptilian not mammalian. The individual becomes incredibly temperature sensitive and temperature intolerant.

The shut off of metabolism also leads to a profound sense of weakness. This weakness can be experienced as a sense of not feeling well, just having the life force sucked out of you, having no strength, or becoming cognitively or judgmental impaired. The end result is that you can not generate energy on demand because your systems are shut down. One becomes weak and fatigued for no obvious reason. You literally lack the strength to do any activity either mental or physical. At the same time the individual may notice an abrupt change in their temperature sensitivity and their ability to tolerate abrupt changes in temperature even of only a few degrees. These dynamic changes reflect the individual's experience. They are the hallmark of this condition. Sadly and tragically many individuals may be misdiagnosed solely because clinical medicine lacks the clinical models to match these symptoms. Note that most doctors are set up to play a matching game. If the symptoms (what the patient is complaining of) or the clinical findings match an existing syndrome you get a diagnosis and treatment. If there is no match than the patient many times will be written off, despite the fact that it is clear to the patient or their family members that the person is ill and incapacitated. Most of the clinical profiles that are referred to as standard blood tests, do not even measure the basic anti-oxidants that have been shown, as indicators of oxidative damage to the body. In many cases the doctor will reassure the patient in the absence of doing any testing. The absurdity of this situation is that the doctor, who is presented as professional vested in using clinical science to validate an illness, rests his diagnosis on opinion not science.

9. Cholesterol

Studies have shown that the oxidative and direct burn damage can have a profound effect on other measures that the doctor often looks at. Cholesterol levels have been found to be elevated after direct microwave exposure.

10. Aging

Protein metabolism can be profoundly affected by microwave exposure. Proteins make up the backbone of fibers in cells which contribute to the maintenance, shape and proper functioning of the cell. Fibers are critical in allowing proper cell division to occur. If this process is interfered cancer can result as well as defects in genetic repair. Also protein strands within the cell help to maintain the proper alignment of structures which are critical to maintaining the integrity of the cell itself. The body is a master weaver. The orderly process or the disruption of this process can result in the failure of cell division, mutations and altered protein production which is critical in manufacturing, enzymes, antibodies, structural proteins which are critical to maintaining cell vitality.

One example of extreme aging that is seen in clinical practice but which is very rare is a condition called Progeria. These are the shriveled up old men who in reality are only teenagers. These are individuals who age prematurely at an extremely fast rate. One of the defects that they seem to have is a failure to produce conectin a fine protein that acts as a shock absorber inside of every cell. One theory holds that their cells age prematurely because they can not absorb shock damage which everyone seems to encounter, but to which they are extremely susceptible.

Additional studies have looked at the levels of anti-oxidants within the cell as markers of increased susceptibility to aging. One consistent finding is that a drop in the level of catalase in the mitochondria or SOD is noted to be associated with a decreased life span. These anti-oxidants are normally present and help to maintain these two all important structures of the cell. Refer to the reference section at the end of this chapter and in the last chapter of this book for more information on their roles.

11. Cancer

Microwave radiation has been shown to be an independent risk factor for developing cancer. Also microwave radiation has been shown to increase ones risk to developing cancer from what ever the cause. Microwave radiation has a permissive effect on cancer promotion from petrochemicals and pollutants. See the reference section at the end of this chapter. In addition pre-malignant cells are extremely vulnerable to viral infections which have been shown in multiple studies to be cancer promoting.

Brain cancer, lymphomas, and cancers of the generative organs have been shown to be induced by microwave radiation. There was a study done that linked increased testicular cancer among highway patrolmen. The officers had a habit of resting the gun on their laps while it was still on.

The bone marrow is critically important in manufacturing blood cells. An impairment of this function can result in anemia and also in an increase in leukemia. Leukemia is a malignancy affecting the white blood cells which are formed in the bone marrow. There seems to be a sharp rise in the incidence of childhood leukemia and lymphomas which have been reported to be caused by microwave radiation.

12. Cystic diseases

There is a rise in cystic disease which has partially been attributed to toxic chemicals, hormonal factors, or genetic disorders which may be attributable to microwave radiation. These illnesses are characterized by the formation of cysts, fluid filled sacs within organs, some which are pre-malignant.

Cysts are similar to blisters within body tissues. When you get a second degree burn of the skin the skin forms a blister. The disruption of blood flow causes the area affected to fill with fluid. In a similar manner burn damage within the body will result in fluid filled sacs called cysts.

Microwave radiation is one of the few modalities that have the ability to penetrate deep into body tissues. Where the microwave waves cross there is an increased tendency to heating and disruption of tissues. This is the same effect that one relies on to cook a large piece of meat within a microwave oven!

Parts of the body affected can include the brain and spinal cord. Cystic changes are noted in multiple sclerosis, abscesses, benign cysts, and hydrocephalus. There is a rise in thyroid cysts. Breast cysts are increasing. There is an increase in cystic changes which occurs in the solid organs of the body which includes the spleen, kidneys, liver, pancreas, gonads and bone marrow. One of the common terms employed is multi-cystic or polycystic disease.

All of these illnesses have the same appearance and probably reflect a common cause. The point is that the pattern of injury or disruption is the same, which reflects a common etiology. The pattern of injury is consistent with microwave radiation injury. Ask yourself if this is mere coincidence?

Protein Synthesis:

A. CONNECTIVE TISSUE

Protein synthesis can be profoundly affected by microwave radiation. The body is a master weaver. The body weaves broad strands out of protein which are critical in maintaining the shape and function of many tissues in the body. Collagen is one linear protein which forms the matrix of many tissues. Collagen is found in the joints and muscles. Collagen is also a key component in the formation of bone; it helps to maintain the elasticity of blood vessels, connective tissues and the skin. It provides for the scaffolding that holds our muscles and bones together. Microwave radiation will contribute to osteoporosis because bone cells need collagen in order to store minerals.

B. IMMUNE FUNCTION

Microwave radiation can disrupt protein production in general. Proteins make up the major constituents of the immune system. Imagine if one had a police or fire force without the equipment to do their job. They would be relegated to a purely symbolic role, but they would not be capable of functioning in any significant way. In a similar fashion if the immune system is compromised then it too could not do its job. What would be the consequence? Well an individual would have an increased risk of and number of infections in various parts of their body. They would be continuously sick and prone to the most devastating infections.

C. Enzyme function

Proteins make up a major part of the nervous system. They are critical in maintaining traffic control within the nervous system, providing transport for critical proteins involved in supporting the endocrine system. Additionally proteins are pivotal in maintaining the production of enzymes which are critical in breaking down our foods and also in breaking down and repairing tissues within the body. Impairment in this area could present in a multitude of ways. One would be a state of semi-starvation along with an increased appetite. The absence or decline in the amount of enzymes suggests that one may shift from absorbing ten cents of every dollar of food eaten to five cents. Some individuals who are semi-starved and eating highly processed foods will tend to go into a hibernation mood of survival. That is they will tend to store fat in the body. Also the ability of microwave radiation to shut down ATP generation will have several effects on the overall economy of the body. If ATP production is limited the body will have a difficult time storing sugars as glycogen. This will affect the overall function of the liver. The liver is pivotal in handling the balance of fats in the body. Instead of excreting fats the balance is shifted towards storing fats and burning protein. Cholesterol levels tend to go up when there is increased oxidative stress on the body. Also cholesterol levels tend to rise when the metabolism is stressed and unbalanced. Since sugars can not be stored the body goes into a mode of storing fat or converting sugars into fat. The net result is that protein production decreases in the body, homocysteine levels elevate, cholesterol levels elevate, metabolism plummets, the liver becomes fatty and the individual becomes unbalanced. Sound familiar. Obesity can be another marker of stress and poor diet affecting the body. In either case the decrease in the availability of digestive enzymes will tend to decrease the amount of food that the individual can process. Since microwave radiation has the ability to shut down protein production the inability to maintain digestive enzymes will have a devastating effect on the body. The net effect if not corrected is a downward and precipitous decline in the overall health of the individual.

D. TISSUE REPAIR/ACCELERATED AGING

The failure to produce or maintain proteins would result in a noticeable inability to repair tissues or maintain them. This is politely referred to as inappropriate or accelerated aging. Accelerated aging or poor wound healing could present in a variety of ways. One might notice a decrease in the thickness and elasticity of the skin. Secondarily this could also present as decreased growth of protein related structures, such as hair or nails. Decrease inability to repair within the body could present as increase in joint problems, advance in arteriosclerosis or blood vessel damage which could initially present as high blood pressure. The nuts and bolts of the current homocysteine model used as a measure of the breakdown of blood vessels simply reflect a breakdown in maintenance of blood vessel structural proteins. The common feature to all these imbalances is the inability to maintain the integrity of protein repair.

Proteins are central to maintaining the integrity of the body. Indeed European studies especially work done by Russian scientist in the last 30 years have developed simple enzyme assays which are reliable and reproducible as markers of microwave radiation damage.

E. FLUID BALANCE

Protein balance is critical to maintain fluid balance in the body. If there is a severe decrease in proteins than fluid seeps into the tissue spaces and produces what is called edema.

14. Symptoms

The point that I made earlier was to focus on the pattern of an illness not the term or name that the doctor refers to it by. Regardless of what the doctor calls it the pattern of burn injury and response of the body is the same. The pattern looks the same:

1. Blood flow diminishes

2. Organ function deteriorates

3. The body breaks down

4. The body starts to loose the ability to repair or heal itself

5. Body metabolism is profoundly affected with the inability to stay warm in a cold climate or the inability to cool off in a warm climate.

6. The individual suffers from profound fatigue

7. The individual can note a prompt drop-off in functionality, for example a decrease in eyesight or mental acuity that starts abruptly without any discernable cause.

8. Frequent and unremitting colds, infections or not feeling well all the time, associate with muscle aches, arthralgias and other manifestations.

The bottom line is with microwave radiation exposure is that your body is being literally slow cooked. All tissues can be affected and will display all or part of the generalized patterns that I stated above.

In terms of the point of this book, one should ask oneself, how could everyone across different regions of the country be getting sick at the same time? Also how is it that the number of ill people seems to be increasing in numbers!

Focus on the common link. There are not many things in nature that can affect individuals across broad geographic areas.

Microwave radiation exposure is the singular unifying link till proven otherwise. There should be no debate. Apart from all the other environmental toxins and pollutants that we are exposed to microwave radiation will act as accelerant of illness.

One may ask how does one gauge this rise in illness?

The simplest way is to look at the increase in medical claims and diagnosis in any period. These numbers are provided primarily through Medicare and Medicaid and are secondarily used by most health insurance companies. If the rate of increase of any particular illness is rising at the same rate across broad geographic areas this is an indicator that there is a single element affecting all those regions equally. I will demonstrate this point in a subsequent chapter.

Reference Section:

OXIDATIVE DAMAGE

Science. 2005 Jun 24;308(5730):1909-11. Epub 2005 May 5 **Extension of murine life span by overexpression of catalase targeted to mitochondria. Schriner SE, Linford NJ, Martin GM, Treuting P, Ogburn CE, Emond M, Coskun PE, Ladiges W, Wolf N, Van Remmen H, Wallace DC, Rabinovitch PS.**
Department of Genome Sciences, University of Washington, Seattle, WA 91895, USA.
Median and maximum life spans were maximally increased (averages of 5 months and 5.5 months, respectively) in MCAT animals. Cardiac pathology and cataract development were delayed, oxidative damage was reduced, H2O2 production and H2O2-induced aconitase inactivation were attenuated, and the development of mitochondrial deletions was reduced. These results support the free radical theory of aging.

Physiol Genomics. 2003 Dec 16;16(1):29-37 **Life-long reduction in MnSOD activity results in increased DNA damage and higher incidence of cancer but does not accelerate aging. Van Remmen H, Ikeno Y, Hamilton M, Pahlavani M, Wolf N, Thorpe SR, Alderson NL, Baynes JW, Epstein CJ, Huang TT, Nelson J, Strong R, Richardson A.**
Lowered levels of SOD are associated with higher levels of cancer and genetic damage.

FASEB J. 2000 Feb;14(2):312-8 **Oxidative damage to mitochondrial DNA is inversely related to maximum life span in the heart and brain of mammals. Barja G, Herrero A.**
Department of Animal Biology-II (Animal Physiology), Faculty of Biology, Complutense University, Madrid 28040, Spain.

Breast Cancer Res Treat. 2000 Jan; 59(2):163-70 **Lipid peroxidation, free radical production and antioxidant status in breast cancer.**

Ray G, **Batra S**, **Shukla NK**, **Deo S**, **Raina V**, **Ashok S**, **Husain SA**.
Department of Biosciences, Jamia Millia Islamia, New Delhi, India.
Reactive oxygen metabolites (ROMs), including superoxide anion (O_2^*-), hydrogen peroxide (H_2O_2) and hydroxyl radical (*OH), play an important role in carcinogenesis. There are some primary antioxidants such as superoxide dismutase (SOD), glutathione peroxidase (GPx) and catalase (CAT) which protect against cellular and molecular damage caused by the ROMs. We conducted the present study to determine the rate of O_2^*- and H_2O_2 production, and concentration of malondialdehyde (MDA), as an index of lipid peroxidation

Clin Chim Acta. 2001 Mar;305(1-2):75-80
Antioxidant enzyme activities and malondialdehyde levels related to aging. Inal ME, Kanbak G, Sunal E.
Osmangazi University, The Medical School, Department of Biochemistry, 26480, Eskisehir, Turkey. minal@ogu.edu.tr
BACKGROUND: Free oxygen radicals have been proposed as important causative agents of aging. We have evaluated age-related changes in antioxidant enzyme activities and lipid peroxidation. METHODS: We measured erythrocyte superoxide dismutase (SOD), catalase (CAT), glutathione peroxidase (GPx) and plasma malondialdehyde (MDA) levels. O

BRAIN FUNCTION

J Sleep Res. 2002 Dec;11(4):289-95
Electromagnetic fields, such as those from mobile phones, alter regional cerebral blood flow and sleep and waking EEG.
Huber R, **Treyer V**, **Borbely AA**, **Schuderer J**, **Gottselig JM**, **Landolt HP**, **Werth E**, **Berthold T**, **Kuster N**, **Buck A**, **Achermann P**.
Institute of Pharmacology and Toxicology, University of Zurich, Zurich, Switzerland

Microwave lens effects in humans. Arch Ophthalmol. 1972 Sep;88(3):259-62.
PMID: 5053251 [PubMed - indexed for MEDLINE

Lenticular and retinal changes secondary to microwave exposure. Acta Ophthalmol (Copenh). 1973;51(6):764-71. .
This article discuses the changes microwave radiation produces to the eye.

Microwave cataract in radio-linemen.
Lancet. 1984 Sep 29;2(8405):760. .
This article discusses the increased incidence of cataracts found in radiolineman.

Invest Ophthalmol Vis Sci. 1977 Apr;16(4):315-9 **Effects of repeated microwave irradiations to the albino rabbit eye. Hirsch SE, Appleton B, Fine BS, Brown PV.**

It appears that the nonprogressive posterior cortical cataracts are a result of the temperature levels generated by the microwaves in the immediate retrolental area. **Animals exposed to microwave radiation had a greater incidence of cataracts.**

Antioxidant enzymes in RBCs as a biological index of age related macular degeneration.
Prashar S, Pandav SS, Gupta A, Nath R. Department of Biochemistry, Postgraduate Institute of Medical Education and Research, Chandigarh, India.
The present study was undertaken to assess the levels of antioxidant enzymes in red blood cells of subjects with age-related macular degeneration and age-matched controls. The results obtained show a significant decrease in activities of superoxide dismutase (p < 0.001) and glutathione peroxidase (p < 0.001) as compared to the controls **Patients with established macular degeneration had substantially decreased levels of SOD and glutathione.**

Doc Ophthalmol. 2003 Mar; 106(2):129-36   **Nitric oxide and lipid peroxidation are increased and associated with decreased antioxidant enzyme activities in patients with age-related macular degeneration.**   Ereklioglu C, Er H, Doganay S, Cekmen M, Turkoz Y, Otlu B, Ozerol E.
Department of Ophthalmology, Erciyes University Medical Faculty, Kayseri, Turkey. evereklioglu@hotmail.com
BACKGROUND: Nitric oxide (NO), hydroxyl radical (OH*), superoxide anion (O2-) and hydrogen peroxide (H2O2) are free-radicals released in oxidative stress. Superoxide dismutase (SOD), glutathione peroxidase (GSHPx) and catalase (CAT) are antioxidant enzymes, mediating defense against oxidative stress. Excess NO and/or defective antioxidants cause lipid peroxidation, cellular dysfunction and death. Age-related maculopathy (ARM) or degeneration (ARMD) is the leading cause of irreversible blindness in developed countries   **Patients with macular degeneration where noted to have decreased levels of protective antioxidants; SOD, catalase, and glutathione. Additionally they had increased evidence of destruction to fats (lipids) in the body..**

Retina. 1993;13(3):230-3.   **Visual abnormalities associated with high-energy microwave exposure.**
Lim JI, Fine SL, Kues HA, Johnson MA   Department of Ophthalmology, Wilmer Ophthalmological Institute, Johns Hopkins Hospital, Baltimore, Maryland. An individual was exposed to microwave radiation, he noted a foreign body sensation and blurring of vision. Ophthalmoscopic examination showed bilateral, small hard drusen. Ancillary tests were consistent with abnormal cone function.Electroretinogram testing revealed a marked decrease in the flicker electroretinogram   **An individual exposed to microwave radiation had evidence of burn damage to the eye as decreased vision as well as other abnormalities. .**

Wei Sheng Yan Jiu. 1999 Jul;28(4):200-2   **[Lipid peroxide damage in retinal ganglion cells induced by microwave]   Yang R, Chen J, Liu X.**
Department of Military Hygiene, Fourth Military Medical University, Xi'an 710032.
The determination of lipid peroxide damage in the primary cultured pig retinal ganglion cells induced by microwave can provide some

experience on the effect of microwave and the protection from its damage. Retinal ganglion cells were cultured in vitro and exposed to different intensities or time of microwave, and cultured for another 48 hours. The content of superoxide dismutase (SOD) and malondialdehyde (MDA) was detected. Results showed that the content of MDA was increased and SOD decreased after radiation. **Direct microwave radiation to retinal cells resulted in decreased SOD levels as well as increased evidence of damage to the cell membranes.(lipid perioxidation).**

BEHAVIORAL AND COGNITIVE DISTURBANCES

The following articles show the relationship between the type of brain wave activity which reflects regional blood flow, the attempts to control brain wave activity and its beneficial effect on syndromes associated with anxiety. Note I have included a study involving alcoholic withdrawal. In addition the last articles in the series document the repetitive nature of diminished blood flow in all of these different syndromes.

Clin Neurophysiol. 2004 Nov;115(11):2452-60
The effects of neurofeedback training on the spectral topography of the electroencephalogram.
Egner T, **Zech TF**, **Gruzelier JH**.Division of Neuroscience and Psychological Medicine, Faculty of Medicine, Imperial College London, UK. tegner@fmri.columbia.edu
The association between alpha/theta training and replicable reductions in frontal beta activity constitutes novel empirical neurophysiologic evidence supporting inter alia the training's purported role in reducing agitation and anxiety **Biofeedback training reduced the incidence of agitation and anxiety.**

Alcohol Clin Exp Res. 1992 Jun;16(3):547-52
Alterations in EEG amplitude, personality factors, and brain electrical mapping after alpha-theta brainwave training: a controlled case study of an alcoholic in recovery.
Fahrion SL, **Walters ED**, **Coyne L**, **Allen T**.

Menninger Clinic, Topeka, KS 66601.
Training consisted of six sessions of thermal biofeedback to increase central nervous system (CNS) relaxation. Effects were documented with pretreatment and post-treatment.post-test the brain map indicated pain-associated EEG activity in the contra lateral somatosensory area, but no apparent anxiety-associated EEG activity. At 4 months post-treatment the patient's wife and colleagues report the patient appears to function in a more relaxed way under the impact of stress, and he reports no longer experiencing craving for alcohol. **Biofeedback training helped an alcoholic to stop drinking and decrease stress levels.**

J Clin Psychol. 1995 Sep;51(5):685-93 **Alpha-theta brainwave neurofeedback training: an effective treatment for male and female alcoholics with depressive symptoms.**
Saxby E, Peniston EG.
Biofeedback Center, Pacific Grove, CA 93950, USA
On the Millon Clinical Multiaxial Inventory-I, the experimental subjects showed significant decreases on the BR scores: schizoid, avoidant, dependent, histrionic, passive-aggression, schizotypal, borderline, anxiety, somatoform, hypomanic, dysthmic, alcohol abuse, drug abuse, psychotic thinking, and psychotic depression. Twenty-one-month follow-up data indicated sustained prevention of relapse in alcoholics who completed BWNT. **Biofeedback helped severe alcoholics who were depressed and exhibited severe personality disorders to stop drinking and helped to alleviate their depression.**

Biol Psychiatry. 1986 Aug;21(10):889-99 **Pathological cerebral blood flow during motor function in schizophrenic and endogenous depressed patients.**
Guenther W, Moser E, Mueller-Spahn F, von Oefele K, Buell U, Hippius H.In Type II schizophrenics and severely endogenous depressed patients, however, we found a widespread nonreactivity of the regional cerebral blood flow (rCBF) to motor activation, with no flow increase in the contralateral primary motor area. **Severely**

depressed and schizophrenic patients where noted to have diminished blood flow to the left side of the brain.

Psychiatry Res. 1996 Nov 25;68(1):1-10 **Temporal lobe dysfunction and correlation of regional cerebral blood flow abnormalities with psychopathology in schizophrenia and major depression--a study with single photon emission computed tomography.**
Klemm E, **Danos P**, **Grunwald F**, **Kasper S**, **Moller HJ**, **Biersack HJ**.
Department of Nuclear Medicine, University of Bonn, GermanyOur data suggest that left-sided temporal lobe dysfunction is related both to schizophrenia and major depression. The localization of hypo perfusion seems to be associated with the type of psychopathology (positive vs. negative symptoms in schizophrenia). Thus, the results support the model of Para limbic and prefrontal dysfunction in both diseases.   **The left temporal lobe appeared to have decreased blood flow in schizophrenic and depressed patients. This was a consistent finding.**

Schizophr Res. 1997 Oct 30;27(2-3):105-17 **Regional cerebral blood flow in late-onset schizophrenia: a SPECT study using 99mTc-HMPAO.**
Sachdev P, **Brodaty H**, **Rose N**, **Haindl W**.
School of Psychiatry, University of New South Wales, Little Bay, Australia.
The LOS(late onset schizophrenic) subjects had a significantly lower cerebral hemispheric perfusion than controls, with a lower perfusion in the frontal and temporal lobes bilaterally. The LOS group also had significantly lower left-to-right hemisphere blood flow ratios. **Older schizophrenic patients where noted to have decreased blood flow to the front of their brains more marked on the left.**

Arch Gen Psychiatry. 1999 Dec;56(12):1117-23
Functional imaging of memory retrieval in deficit vs. nondeficit schizophrenia.

Heckers S, **Goff D**, **Schacter DL**, **Savage CR**, **Fischman AJ**, **Alpert NM**, **Rauch SL**.Psychotic Disorders Unit, Psychiatric Neuroimaging Research Group, Massachusetts General Hospital, Boston, USA. heckers@psych.mgh.harvard.edu

During the attempt to retrieve poorly encoded words, patients without the deficit syndrome recruited the left frontal cortex to a significantly greater degree than did patients with the deficit syndrome. The 2 schizophrenia subtypes did not differ in the activity or recruitment of the hippocampus during memory retrieval. CONCLUSION: Frontal cortex function during memory retrieval is differentially impaired in deficit and nondeficit schizophrenia, whereas hippocampal recruitment deficits are not significantly different between the 2 schizophrenia groups. **Schizophrenic patients of both types had decreased ability to use their left frontal lobes, they were impaired in memory tasks.**

J Neurol Neurosurg Psychiatry. 1997 Nov;63(5):597-604
Temporal lobe abnormalities in dementia and depression: a study using high resolution single photon emission tomography and magnetic resonance imaging.
Ebmeier KP, **Prentice N**, **Ryman A**, **Halloran E**, **Rimmington JE**, **Best JK**, **Goodwin GM**.
MRC Brain Metabolism Unit, Royal Edinburgh Hospital, Morningside Park, UK.
Demented patients showed reduced perfusion, particularly in the left temporoparietal cortex. In these regions of interest, patients with late onset depression tended to have perfusion values intermediate between patients with early onset depression and demented patients. **Patients with late onset dementia and patients with depression both showed patterns of decreased blood flow to the left frontal lobes.**

Arch Gen Psychiatry. 1994 Sep;51(9):677-86
Reduction of cerebral blood flow in older depressed patients.
Lesser IM, **Mena I**, **Boone KB**, **Miller BL**, **Mehringer CM**, **Wohl M**.

Department of Psychiatry, Harbor-UCLA Medical Center, Torrance.
Patients exhibited a global reduction in regional cerebral blood flow compared with controls, with the orbital frontal and inferior temporal areas affected bilaterally. **Older depressed patients had a significant decrease in the blood flow to their frontal lobes.**

Nucl Med Commun. 2005 Sep;26(9):757-63
Voxel-based assessment of spinal tap test-induced regional cerebral blood flow changes in normal pressure hydrocephalus.
Dumarey NE, **Massager N**, **Laureys S**, **Goldman S**.
aDepartment of Nuclear Medicine and PET/Biomedical Cyclotron Unit bDepartment of Neurosurgery, Hopital Erasme, Universite Libre de Bruxelles, Brussels, Belgium cCyclotron Research Centre, Universite de Liege, Liege, Belgium.
According to SPM analysis, gait improvement at the spinal tap test in patients with NPH syndrome is associated with an rCBF increase localized in the bilateral dorsolateral frontal and left mesiotemporal cortex **Older patients with NPH (which is a swollen brain) and presented with impaired ability to walk, improved after decompression of the brain. There was noted to be an improvement which paralleled a return to normal pattern of blood flow in the brain. This showed that the nervous system was intact, it was just suffering from decreased blood flow.**

J Nucl Med. 1999 Feb;40(2):244-9.
Parametric mapping of cerebral blood flow deficits in Alzheimer's disease: a SPECT study using HMPAO and image standardization technique.
Imran MB, **Kawashima R**, **Awata S**, **Sato K**, **Kinomura S**, **Ono S**, **Yoshioka S**, **Sato M**, **Fukuda H**.
Department of Nuclear Medicine and Radiology, Aoba Brain Research Imaging Center, Institute of Development, Aging and Cancer, Tohoku University, Tohoku University Hospital, Sendai, Japan

The frontal regions of the brain, in addition to parietal and temporal lobes, may show reduced CBF(blood flow) in patients with AD(Alzheimer's disease) even at an early stage of dementia. **Patients with early onset Alzheimer's disease showed decreased blood flow to the frontal lobes of the brain.**

Bioelectromagnetics. 1998;19(6):384-7
Effects of microwaves emitted by cellular phones on human slow brain potentials.
Freude G, Ullsperger P, Eggert S, Ruppe I.
Federal Institute for Occupational Safety and Health, Berlin, Germany. freude@baua.de   EMF exposure effected a significant decrease of preparatory slow brain potentials SPs at central and temporo-parieto-occipital brain regions.   **Microwave radiation decreased psp in the brain.**

Eur J Appl Physiol. 2000 Jan;81(1-2):18-27
Microwaves emitted by cellular telephones affect human slow brain potentials.
Freude G, **Ullsperger P**, **Eggert S**, **Ruppe I**.
Bundesanstalt fur Arbeitsschutz und Arbeitsmedizin, Noeldnerstrasse 40-42, D-10317 Berlin, Germany. freude@baua.de
The influence of electromagnetic fields (EMF) emitted by cellular telephones on preparatory slow brain potentials (SP) was studied in two experiments, about 6 months apart. In the first experiment, a significant decrease of SP was found during exposure to EMF in a complex visual monitoring task (VMT).
Microwave radiation reduced the psp waves in two repeat experiments.

NEUROTRANSMITTER DYSFUNCTION

[5-HT contents change in peripheral blood of workers exposed to microwave and high frequency radiation]
Zhonghua Yu Fang Yi Xue Za Zhi. 1989 Jul; 23(4):207-10. Chinese.
The 5-HT contents in the whole blood were inversely proportional to the power density of microwave and high frequency groups. The incidence of neurasthenic syndrome, unsymmetrical skin

temperature in both limbs and hypotension was higher in the microwave and high frequency radiation groups than that in the control group. The incidence of bradycardia and some abnormal items of electrocardiograph in the microwave group was obviously greater.

Note decreased 5-HT levels are associated with decreased serotonin levels and are an indicator of inflammation. 5-HT is the direct source in the body of serotonin. Decreased levels of serotonin are associated with depression or anxiety regardless of the illness that they may associated with, that is decreased serotonin levels are associated with cancer and depression, depression and rheumatic illness, depression and neurological disease, depression and infectious disease.

Nervous and behavioral effects of microwave radiation in humans. Am J Epidemiol. 1973 Apr; 97(4):219-24. Review.

Int Clin Psychopharmacol. 2004 Mar;19(2):89-95 **Antioxidant enzyme activities and oxidative stress in affective disorders.**
Ozcan ME, Gulec M, Ozerol E, Polat R, Akyol O.
Department of Psychiatry, Inonu University Medical School, Malatya, Turkey. eozcan@inonu.edu.tr
Recent data from several reports indicate that free radicals are involved in the biochemical mechanisms underlying neuropsychiatric disorders in human. The results of several reports suggest that lower antioxidant defences against lipid peroxidation exist in patients with depression and that there is a therapeutic benefit from antioxidant supplementation in unstable manic-depressive patients.
There was found to be lowered anti-oxidant levels in depression and mania. Supplementation tended to improve the clinical situation.

Med Pr. 2002;53(4):311-4 **[Effect of electromagnetic field produced by mobile phones on the activity of superoxide dismutase (SOD-1) and the level of malonyldialdehyde (MDA)--in vitro study]** Stopczyk D, Gnitecki W, Buczynski A, Markuszewski L, Buczynski J.
Zakladu Medycyny Zapobiegawczej i Promocji Zdrowia, Wojskowej Akademii Medycznej w Lodzi. darstop@poczta.onet.pl

The aim of the study was to assess in vitro the effect of electromagnetic field produced by mobile phones on the activity of superoxide dismutase (SOD-1) and the level of malonyldialdehyde (MDA) in human blood platelets. The suspension of blood platelets was exposed to the electromagnetic field with the frequency of 900 MHz for 1, 3, 5, and 7 min. Our studies demonstrated that microwaves produced by mobile phones significantly depleted SOD-1 activity after 1, 5, and 7 min of exposure and increased after 3 min in comparison with the control test. There was a significant increase in the concentration of MDA after 1, 5, and 7 min and decrease after 3 min of exposure as compared with the control test. **Even after one minute of exposure to cell phones, anti-oxidant levels in the blood had dropped significantly.**

Environ Health Perspect. 2004 May; 112(6):687-94 **Magnetic-field-induced DNA strand breaks in brain cells of the rat.** Lai H, Singh NP.
Bioelectromagnetics Research Laboratory, Department of Bioengineering, University of Washington, Seattle, Washington 98195-7962, USA. hlai@u.washington.edu **Microwave produces DNA damage in brain cells.**

Bioelectromagnetics. 1995; 16(3):207-1 **Acute low-intensity microwave exposure increases DNA single-strand breaks in rat brain cells.** , Singh NP.
Department of Pharmacology, University of Washington, Seattle 98195, USA.
Microwaves produce DNA damage in brain cells. Pharmacol Biochem Behav. 1989 May; 33(1):131-8. **Low-level microwave**

irradiation and central cholinergic systems. <u>Lai H</u>, <u>Carino MA</u>, <u>Horita A</u>, <u>Guy AW</u>.

Department of Pharmacology, University of Washington School of Medicine, Seattle 98195.

J Neurochem. 1987 Jan; 48(1):40-5 **Low-level microwave irradiations affect central cholinergic activity in the rat.** <u>Lai H</u>, <u>Horita A</u>, <u>Chou CK</u>, <u>Guy AW</u>.
Pulsed microwave irradiation (2-microseconds pulses, 500 pulses/ s) decreased choline uptake in the hippocampus and frontal cortex but had no significant effect on the hypothalamus. **Microwave radiation decreased the uptake of choline in the frontal lobes and hippocampus(memory circuit).**

Bioelectromagnetics. 1992;13(1):57-66 **Single vs. repeated microwave exposure: effects on benzodiazepine receptors in the brain of the rat.** <u>Lai H</u>, <u>Carino MA</u>, <u>Horita A</u>, <u>Guy AW</u>.

Department of Pharmacology, University of Washington School of Medicine, Seattle 98195.

Benzodiazepine receptors in the brain are responsive to anxiety and stress, our data support the hypothesis that low-intensity microwave irradiation can be a source of stress. **Microwave radiation activated benzodiazepine(valium) receptors in the brain.**

Radiobiologiia. 1991 Mar-Apr;31(2):257-60 [**Action of microwaves with different modulation frequencies and exposure times on GABA receptor concentration in the cerebral cortex of rat.** <u>Kuznetsov VI</u>, <u>Iurinskaia MM</u>, <u>Kolomytkin OV</u>, <u>Akoev IG</u>.
The effect of 800 mHz microwaves of 0, 3, 5, 7, 16, and 30 Hz modulation on GABA receptor concentration in rat brain cortex has been investigated. Irradiation of the whole body at a modulation frequency of 16 Hz readily decreases the GABA receptor concentration. **Microwave radiation decreased the amount of GABA (calming neurotransmitter) in the brain.**

ENERGY METABOLISM

Bioelectromagnetics. 1980;1(2):171-81
Microwave effects on energy metabolism of rat brain.
Sanders AP, Schaefer DJ, Joines WT
The data support the hypothesis that microwave exposure inhibits mitochondrial electron transport chain function, which results in decreased ATP and CP (creatine phosphate) levels in brain.
Note: that many of the common anti-oxidants of the body compromise the functioning parts of the mitochondrial electron transport chain and operate in a dual function as a reservoir of anti-oxidants and to preserve membrane function.

Bioelectromagnetics. 1985;6(1):89-97
Effects of continuous-wave, pulsed, and sinusoidal-amplitude-modulated microwaves on brain energy metabolism.
Sanders AP, Joines WT, Allis JW.
Since brain temperature did not increase, the microwave-induced increase in brain NADH and decrease in ATP and CP concentrations was not due to hyperthermia. This suggests a direct interaction mechanism and is consistent with the hypothesis of microwave inhibition of mitochondrial electron transport chain function of ATP production **Microwave radiation directly affects the mitochondria(cell furnaces) causing a decrease in ATP and creatine phosphate, both necessary components of energy production.**

Radiat Res. 1985 Dec;104(3):365-86
The relationship of decreased serum thyrotropin and increased colonic temperature in rats exposed to microwaves.
Lu ST, Lebda NA, Pettit S, Michaelson SM
Results showed that TSH concentration decreased in rats after microwave exposure.
Decreased TSH concentration was usually accompanied by increased colonic temperature.

Note: that TSH stimulates the thyroid gland to make thyroid hormone. If the levels are going down thyroid production is being shut down.

Radiat Res. 1986 Aug;107(2):234-49

Effects of microwaves on the adrenal cortex.

Lu ST, Pettit S, Lu SJ, Michaelson SM

Increases in CS (steroid, cortisone) concentration were proportional to power density or colonic temperature and inversely proportional to the baseline CS. Increased CS concentration was never observed without increased colonic temperature and was not persistent 24 h after exposure.

Note: while the thyroid hormones are dropping at the same time the cortisol levels of the body are being elevated in response to increased oxidative stress.

CHOLESTEROL

Int J Radiat Biol. 1999 Jun;75(6):757-66. **Ionizing radiation alters hepatic cholesterol metabolism and plasma lipoproteins in Syrian hamster. Feurgard C, Boehler N, Ferezou J, Serougne C, Aigueperse J, Gourmelon P, Lutton C, Mathe D**. Institut de Protection et de Surete Nucleaire, Departement de Protection de la sante de l'Homme et de Dosimetrie, IPSN, Fontenay-Aux-Roses, France. catherine.feurgard@ipsn.fr

Plasma cholesterol was increased by 77% and triglycerides by +207%. LDL accumulated in plasma while high-density lipoprotein (HDL) levels decreased. HDL contained significant amounts of apo SAA, the acute phase apolipoprotein. Lipoprotein modifications that appeared following radiation exposure may result from an induced inflammatory state and may further contribute to vascular damage

Microwave radiation causes triglycerides to elevate along with an elevation of LDL and a decrease in HDL (good cholesterol). All of these are considered to be indicators of damage to blood vessels.

Neurochem Res. 1988 Jul;13(7):671-7   **Cell membranes: the electromagnetic environment and cancer promotion.   Adey WR**.
VA Medical Center, Loma Linda, California 92357.
Powerful cancer-promoting phorbol esters act at cell membranes to stimulate ornithine decarboxylase which is essential for cell growth and DNA synthesis. This response is enhanced by weak microwave fields, also acting at cell membranes.   **Microwave radiation tends to promote the effect of cancer causing chemicals.**

Environ Health Perspect. 1990 Jun;86:297-305.   **Joint actions of environmental nonionizing electromagnetic fields and chemical pollution in cancer promotion.   Adey WR**.
Research Service (151), Pettis Memorial VA Medical Center, Loma Linda, CA 92357.
From cancer research comes a floodtide of new knowledge about the disruption of communication by cancer-promoting chemicals with an onset of unregulated growth. Bioelectromagnetic research reveals clear evidence of joint actions at cell membranes of chemical cancer promoters and environmental electromagnetic fields. The union of these two disciplines has resulted in the first major new approach to tumor formation in 75 years, directing attention to dysfunctions in inward and outward streams of signals at cell membranes, rather than to damage DNA in cell nuclei, and to synergic actions of chemical pollutants and environmental electromagnetic fields. **Microwave radiation disrupts cellular signaling and tends to promote cancerous changes and acts additively along with chemical pollutants.**

Am Ind Hyg Assoc J. 1993 Apr;54(4):197-204   **Overview of epidemiologic research on electric and magnetic fields and cancer.   Savitz DA**.
Department of Epidemiology, School of Public Health, University of North Carolina, Chapel Hill.

Across many different study designs and settings, certain groups of electrical workers show elevated occurrence of leukemia and brain cancer. The consistency of findings is notable

There was a higher incidence of brain and blood cancer in workers exposed to elevated electromagnectic fields. .

Neurol Res. 1982;4(1-2):115-53   **Nonlinear wave mechanisms in interactions between excitable tissue and electromagnetic fields. Lawrence AF, Adey WR.**

It is now well established that intrinsic electromagnetic fields play a key role in a broad range of tissue functions, including embryonic morphogenesis, wound healing, and information transmission in the nervous system. These same processes may be profoundly influenced by electromagnetic fields induced by an external force. **Microwave radiation or emf can disrupt cell signaling, resulting in a broad range of tissue functions. These can present as poor wound healing, malformations of the fetus(unborn), nervous system disruption.**

J Toxicol Environ Health B Crit Rev. 2004 Sep-Oct;7(5):351-84 **Mobile telephones and cancer--a review of epidemiological evidence.   Kundi M, Mild K, Hardell L, Mattsson MO.**

Institute of Environmental Health, Department for Occupational and Social Hygiene, Medical Faculty, University of Vienna Kinderspitalgasse 15 A-1095 Vienna Austria. Michael.Kundi@univie.ac.at

Overall nine epidemiological studies have been published, four from the United States, two from Sweden, and one each from Denmark, Finland, and Germany. Seven studies were mainly on brain tumors, with one investigating in addition to brain tumors salivary gland cancer and another cancer of the hematopoietic and lymphatic tissues, and one examining intraocular melanoma studies. Nevertheless, all studies approaching reasonable latencies found an increased cancer risk associated with mobile phone use.   **Cell phones, microwave radiation was shown to be correlated with an increase cancer rate of the head and neck in nine independent studies.**

CHAPTER 3:

GEOGRAPHIC INDICATORS

Earlier I had stated that microwave radiation would cause a rise in the incidence of illnesses at the same rate across broad geographic boundaries. What sort of information would reveal this type of trend? Every country does a statistical analysis of illness to observe trends and to help them allocate resources in dealing with health issues.

Within statistical analytic charts there is a reporting of the incidence of disease. This report shows how many individuals per 100,000 have been affected in a given region or by sex, race, economic status, age or whatever other variable one might wish to consider. Every country does this analysis annually to watch out for trends in illness. In the United Kingdom and Ireland there is a breakdown of the incidence of cancer by type, sex and region. There is an excellent publication put out by the British government called the Cancer Atlas of the UK and Ireland. The same information can be obtained by accessing the web address www.stastics.gov.uk.

Within the UK website is a breakdown of cancers by individual types. There are a set of graphs, bar graphs, which show the yearly trend of the incidence of cancers across broad areas of the UK and Ireland. One should note the rate of rise across a particular region. This is the shape of the curve produced by the year to year change in the incidence of cancers. If the curve is similar across broad areas then this suggests that there is an influence that is affecting all regions at the same rate and causing the cancer rate to rise. This can be noted by the upward sloop in any given geographic region. If all regions are showing a similar upward trend independent of the total number of cases than a uniform influence is acting on all of them. Of note is that in the last two year periods, 1997-1999 on the graphs there has been a sharp upswing in the incidence of illness, which parallels the rise in microwave radiation utilization.

For clarification the picture on the cover of this book sums up the concept. We have created an artificial microwave environment and these graphs reflect the changes that are occurring as an outgrowth of that phenomenon.

An analysis of the curves shows that there is an independent increase in the rates of:

1. <u>Brain cancer</u>- has gone up equally across all areas

BRAIN CANCER, INCIDENCE

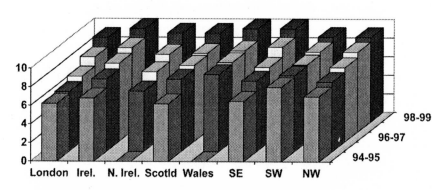

2. <u>Leukemia</u>- have gone up exponentially and equally across all geographic areas.

LEUKEMIA: INCIDENCE PER 100,000

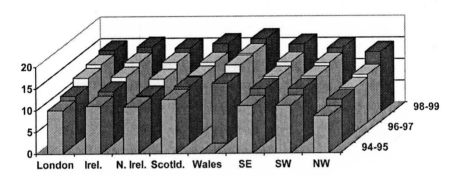

3. <u>Lymphomas</u>-have gone up and show the same rise in incidence across broad geographic areas.

LYMPHOMA, HODGKIN'S DISEASE IN MALES
PER 100,000

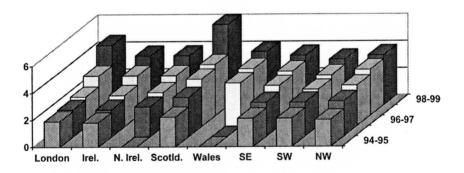

4. <u>Testicular cancer</u> has gone up at an rate increasing exponentially from 1998 to 1999 across all regions

TESTICULAR CANCER: PER 100,000

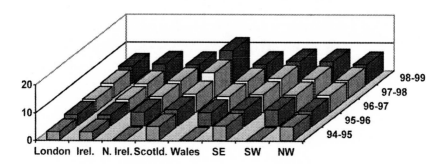

5. <u>Laryngeal cancer</u> has gone up exponentially more so in men at the same rate regionally. Note that laryngeal tissue is one of the most microwave sensitive and can be used as a simple indicator of the rise of illness. This chart does not discount other contributory factors. It reflects that microwave radiation is acting uniformly.

LARYNGEAL CANCER IN MALES, PER 100,000

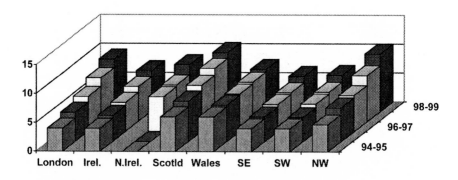

6. <u>Prostate and uterine cancers</u> have shown the same rate change.

PROSTATE CANCER: INCIDENCE PER 100,000

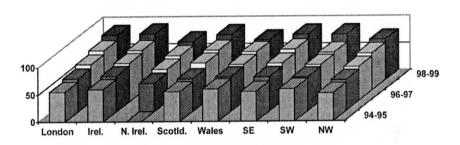

UTERINE CANCER

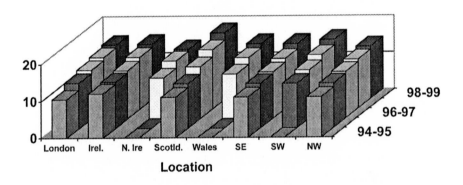

What is interesting about this trend in the UK is that the data fits the dosimetry model and follows a uniform pattern of rise. This rise is independent of geography and parallels the exponential utilization of microwave radiation. .

The data fulfills these three criteria:

1. Matches the dosimetry model

2. Crosses geographic boundaries uniformly

3. Parallels the rise in microwave activity or transmission independently of the method used.

The cancer charts reflect the relationship in the uniform geographic rise in the incidence of cancers to microwave radiation utilization. There has been an exponential increase in other diseases which fall under different names but are linked by the above pattern.

In the next chapter I will provide a sampling of different generalized terms for medical conditions and compare them to the organ systems as displayed in the dosimetry model. The intent of the chapter is to allow the reader the chance to see how a particular disease corresponds to a particular organ system. Bottom line is slow cooking is slow cooking, be it a nerve, muscle or liver cell. Regardless of the many terms that are used to describe illness, they may have a common cause. What is relevant is that the continued upward trend across broad geographic areas reflects the common underlying pathology which is causing disease progression. Most illnesses present with the same generalized features, which include a decrease in function ability, decreased repair rate, decreased ability to perform work and malignant transformation. This is the bottom line of illness, and the main factors of why one seeks medical attention.

Microwave radiation will operate as an independent co-contributory factor that will accelerate the rise in various illnesses across broad geographical areas. An analysis of epidemiological data will demonstrate this pattern.

If one wants to examine the same trends say in the United States, all one has to do is access any epidemiological data base. The simplest to examine would be the rise in the utilization of medical services by diagnosis across all geographic areas. This data is readily accessible. A search utilizing the uniform coding that is used by Medicare and Medicaid to provide for reimbursement by diagnosis would readily reveal the same pattern.

CHAPTER 4:
MEDICAL EQUIVALENTS

The following list includes some common conditions that affect certain organ systems. The list is not exhaustive but does reflect some of the more common illness that present today. The list is designed to help the average reader see the relationship between the dosimetry model and the current diseases which affect the general population. By the same token for anyone who does research, a rise in the incidence of any particular disease across broad geographic areas that seems to have the same rate of change of incidence is probably related to microwave radiation.

The list is not meant to be exhaustive and it does not exclude other factors that may contribute to illness. The point is that microwave radiation can be a powerful direct or co contributory source of illness that may masquerade amongst the many medical terms that are used. Microwave radiation acts indiscriminately affecting all equally, children, men, woman and the elderly. Note the same also applies to pets, domestic and undomesticated animals. For that matter all forms of life. The conditions referred to below apply equally to children as well as adults with obvious exceptions.

Nervous system:

Blood vessel disorders and functional disorders; stroke, cva, headache syndrome, migraine headache, ocular blindness or partial blindness, multiple psychiatric disorders(see previous reference in section 3) depression, anxiety disorders, add, adhd, autism, schizophrenia, alcoholic syndrome, psychosis, senility, Alzheimer's disease, some neuropathies, some balance disorders

Brain cancers, primary, hydrocephalus, MS, any syndrome encompassing cognitive dysfunction along with profound fatigue,pms, light sensitivity disorders, sleep disorders,ALS, peripheral neuropathies, eye disorders including corneal abrasions, cataract, glaucoma, macular degeneration, retinopathies, decreased visual acuity, eye fatigue. Ear disorders including recurrent and repeat ear infections, labyritnithitis, tinnitus, decreased hearing, vertigo

Metabolic or endocrine disorders:

Thyroid disorders, thyroid cyst, thyroid cancer, hypothyroidism, pancreatic cysts, pancreatic dysfunction, metabolic syndrome or syndrome x, adrenal dysfunction, hypoadrenalism, gonadal dysfunction, infertility, decreased libido, decreased hormonal production, obesity, liver dysfunction, Wilson syndrome, metabolic wasting, chronic fatigue syndrome, fibromyalgia, chronic fascitis, syndrome of elevated cholesterol, hypertriglycedemia.

GASTROINTESTINAL DISORDERS;

Recurrent oral infections, tooth and periodontal disease, salivary tumors, pharyngeal (upper throat) infections and tumors, swallowing disorders, esophagitis, esophageal tumors, reflux esophagitis, stomach disorders, stomach ulcers, stomach cancer, lymphoma of the intestinal tract, pancreatic dysfunction, liver disorders, liver cancer, hepatitis, gallbladder dysfunction, malabsorption, constipation and diarrhea

KIDNEY, BLADDER AND SEXUAL FUNCTION

Kidney cysts, decreased kidney function, glomerulonepathies, frequent urination, loss of bladder function(note the nerves, circulation to or the muscles of the bladder can all be involved), bladder tumors, prostate disorders,testiculardisorders,sexualdysfunction,decreased performance and libido, cancers of the genitourinary tract, infertility(male and female), uterine and ovarian cancers, vaginal cancer, recurrent bladder, urinary or vaginal tract infections—note breast cancer and breast cyst are to be included in this list

VASCULAR SYSTEM(ARTERIES, VEINS,LYMPHATICS AND BLOOD FORMING ORGANS:

Arteriosclerosis, vein thrombosis, aneurysms, ischemia, angina, claudication, venous insufficiency, lymphedema, lymphadenopathy, lymphatic cancer, lymph node enlargement, Anemia, bone cancer, leukemia's

Bones and mineral metabolism:

Osteoporosis, osteopenia, demineralization syndrome and mineral malabsorption, arthritis, joint deformities, synovitisis, spinal abnormalities, hip disorders

Immune Dysfunction :

Recurrent infections, Immune dysfunction, persistent and life threatening infections, HIV syndrome, autoimmune disorders (lupus, SLE)

Respiratory System:

Recurrent laryngeal (upper airway) infections, laryngeal cancer, voice abnormalities, bronchitis, recurrent respiratory illnesses, pneumonia, lung cancer

Skin disorders:

Skin cancers. Melanoma, head and neck cancers, squamous cell cancer, basal cell cancers, accelerated skin aging, poor wound healing, recurrent skin infection, recurrent abscesses, psoriasis, eczema, drying skin and hair

CHAPTER 5:

PREVENTIVE STRATEGIES:

The previous chapters established an overview of the health consequences from exposure to microwave radiation, reflecting what has been established in the current scientific and medical literature.

This chapter focuses on supplements which have been shown through the scientific literature to be useful in protecting cells in the body from the effects of microwave radiation. I have focused on simple remedies that the individual can use proactively to protect themselves. With regard to microwave radiation, there are basically two levels of responsibility. The first is the individual and the second is societal. Of the approaches that one could take there are basically two. The first is what natural substances are protective of the body. The second is how a technology can be made safer.

There are already on the marketplace several manufacturers of devices that have been found useful in reducing emf exposure, many of which can be readily accessed by doing a computer database search. Several industries already have set up standards to safe guard their employees. Also manufactures routinely used protective coatings on appliances which are sensitive to microwave transmissions. There is already an industry standard in relation to safe guarding certain appliances, many which depend on maintaining the integrity of microprocessor circuitry.

On the societal level there are several issues which have to be addressed first. As with any problem there is the point of acknowledging that it exists in the first place. That is the purpose of this manuscript. Ultimately the responsibility of affecting change will fall on our legislators and corporations. It is they who will have to come up with innovative strategies for rerouting microwave traffic to minimize its effect on our global population. Perhaps microwave transmissions could be restricted to narrow corridors over the oceans and cabled underground across large continental areas. This might minimized the exposure of the average individual.

As regards supplements I have often encountered several distinct questions that are raised by consumers. I will attempt to address two or three of the more common questions that consumers ask.

1. I eat a good diet why should I have to take supplements?

There are several reasons why relying on the diet alone is not necessarily the best approach. The most obvious point is that if the diet were so good how come you got sick. There are several explanations for this. The first is that the individual's **metabolism is shut down**. The foods may not have the appropriate nutrients. Many of the foods that we eat are highly processed. Additionally many of the active principals that are described in the list are not routinely part of ones regular diet.

The second point is that the overall metabolism of the individual may be shut down. The body's ability to repair and protect itself may be severely limited. **Protein synthesis** and the assembly of essential bionutrients are **shut down**. Natural **repair nutrients** and **anti-oxidants** are being **over utilized** and under replaced. They are being used up faster than they can be replaced. The individual may lack the necessary enzymes to break down nutrients in the body. Enzyme systems may have become inactivated and are sluggish. Major organ systems in the body are relatively inactivated and thus incapable of self correction.

Given these factors often if one begins with simple supplements, one has the opportunity to help oneself and unmask the degree of the problem to begin with. One may question the need for supplements, but in a body that is overwhelmed, that body may not be able to produce these substances fast enough. The amounts it can manufacture are being used up at a rate faster than they can be replaced.

2. Why does not my doctor seem to know about this?

The problem with waiting for one to get permission from ones doctor is several fold.

First medical science is advancing at a faster rate than those advances can be integrated into medical practice. Of the supplements that I listed below and for that matter for most anti-oxidants listed as well, most physicians do not have the blood tests to measure them or for those matters have any awareness beside an opinion as to what appropriate blood levels would be in any clinical situation.

You would have a hard time defining a deficiency disease if you never did a blood test. Simply stated most doctors are not trained in the use of nutrients. The majority of physicians are unaware of them but at the same time do not have either the expertise or laboratory testing to properly evaluate for these deficiencies. As I stated before there is a large lag period before certain testing is even integrated into clinical testing. So even though your physician may state that he could not find anything wrong, did he test for it. A simple example of this is how many individuals go to the doctor and are told they are depressed and put on medication. But they never had a single blood test done that was confirmatory of depression. So they underwent treatment and the outcome of the treatment becomes the basis for the diagnosis. That is the success of the treatment becomes the validation of the diagnosis. If you get better I was right or if you do not then I simple prescribe another treatment.

Strangely enough doctors use this approach all the time because their diagnostic testing abilities are limited. Treatment becomes the basis of diagnosis. However in many cases doctors do not seem to embrace this same approach when it comes to supplementation. A treatment regimen becomes the basis for unmasking a deficiency state. It is quite a double standard.

Medically there is no system set up to recognize or address these issues. The average individual has to become self reliant in taking charge of their own health. The reality is that only the body can repair itself and it uses nutrients to do that job. To use another analogy, regardless of how brilliant the mechanic is the car still depends on gas to run. Do not confuse the two.

3. I am concerned about questioning my doctor, he is the expert what do I know?

Many people feel uncomfortable about questioning their physicians. The patient depends on the doctor for service and recognizes his expertise in certain matters. Many individuals are uncomfortable about offending him or her, appearing stupid or argumentative. Often the individual does not know what to ask of the doctor. Too often the average individual assumes that their doctor has considered the issue or that their health care provider has been educated about the issue. Sadly this is often not the case. The same applies to authorities in the field.

Most only have an opinion. What an individual needs is an empathetic supportive health care provider who is willing to consider the issues and order appropriate tests. At best the physician should be willing to admit that they have no familiarity or expertise in a particular field or area. It is crucial for the patient to understand this point in communicating with their doctor. Often the answer may seem confusing to the patient because the doctor is commenting on someone else opinion. Ask them if they are familiar with the information in this book.

At the same time many fields are too new for the physician to keep up with. It is not that they are stupid or incompetent; it is just that they do not know. The irony of the situation is that many individuals have more information on alternative health and supplements than their doctors.

The following are a list of supplements that have been found to be useful and protective from microwave radiation. They are readily available from several different manufactures and are relatively inexpensive to purchase.

Melatonin

Melatonin is produced naturally in the body. It is not only to induce sleep but has noted to be a powerful anti-oxidant with major anticancer properties. Melatonin has been noted to prevent the breaks in DNA in brain cells which may be the original precursors to cancer and cell death. Additionally melatonin has been found to effective in preventing kidney damage from cell phones. Melatonin is available in many strengths, however there is a low dose form of .5 mg which is simple to take and does not cause adverse effects.

Zinc

Zinc added into the diets of rats was found to be protective in the eye from markers of oxidative damage. Also zinc was found to help preserve the levels of anti-oxidants in the blood and offset damage to cell membranes.

S.O.D

Sod or super oxide dimutase has been found to be a critical enzyme in producing hydrogen peroxide in the body. It levels have been used as markers of oxidative damage from microwaves as well as in other studies of oxidative stress. Sod has been found to be elevated and preserves longevity in animals. Several studies have shown that one of the effects of chronic microwave exposure is decreased life span. This effect seems to be countered by Sod. Also Sod seems to have a more preferential effect in protecting the DNA of mammals. This protective effect seems to be important in reducing mutations in cells which can lead to cell death and cancers.

Gingko Bilboa

Gingko Bilboa is an herb which has been suggested for circulation problems. Recent research shows that gingko is a powerful anti oxidant which prevents oxidative damage in the brain, eye, and kidney. It has been shown to protect the DNA of the cell and minimize cancer production. Additionally gingko has been shown to have potent anti-cancer activities and helps to support the bodies' production of sod, catalase, and glutathione.

Bilberry extract

Bilberry extract has long been known to preserve vision and seems to have a specific action in minimizing fatigue in the retina (back of the eye) and reduce oxidative damage.

Caffeic acid

Is a new compound found in many foods. It has been found to be very important in preventing oxidative damage and preventing DNA damage. Also it protects the kidneys and brain from damage from microwave radiation. It is similar in anti-oxidant properties to melatonin. In some comparative test with polyphenol compounds it has been noted to be vastly superior to polyphenolic compounds. It is found commonly in apple cider, bee propolis and other common foods.

Catalase

Is a necessary enzyme which reduces hydrogen peroxide to water in the body. It regulates the role of peroxide in the body which is used to neutralize many toxins and viral particles. Catalase serves to protect the mitochondria from burn damage. The current notion of illness is that aging is a reflection in the breakdown of mitochondrial functioning. In essence the mitochondrion is your furnace and if it does not work you do not work.

Coq10

A major antioxidant which is found in the respiratory chain. Sufficient coq10 is required to maintain cellular metabolism at the same time that it resists oxidative changes in the cell membrane. Coq has been found useful in minimizing the side effects of severe oxidative stress.

DHA

Along with EPA constitute the class of n3 fatty acids. They have all been found beneficial in decreasing inflammatory changes. They are commonly found in fish oils.

Acetyl-carnitine/R-Lipoic acid:

Is the active form of carnitine which is needed to transport fats through cell membranes. Acetyl-carnitine transports fats to the mitochondria (small metabolic furnace) within the cell so that the body can produce energy. Lipoic acid is an important component of the respiratory chain where it functions with coq10 along with iron and sulfur. Lipoic acid has been shown to be an important antioxidant and additionally it has anti-viral properties.

Lycopene:

Has been shown to be a powerful antioxidant which retards oxidation of fats. Lycopene has been shown to have anti-cancer effects and has been also shown to be protective in prostate disease.

This list is not exhaustive and there are many other resources that one can access to gain further information about this topic.

Reference Section:

MELATONIN

Bioelectromagnetics. 1997;18(6):446-54

Melatonin and a spin-trap compound block radiofrequency electromagnetic radiation-induced DNA strand breaks in rat brain cells.

Lai H, **Singh NP**.
Bioelectromagnetics Research Laboratory, University of Washington, Seattle 98195, USA.

Environ Health Perspect. 2004 May;112(6):687-94 <u>Environ Health Perspect. 2004 Sep;112(13):A726; author reply A726.</u>**Magnetic-field-induced DNA strand breaks in brain cells of the rat.**

Lai H, **Singh NP**.
Bioelectromagnetics Research Laboratory, Department of Bioengineering, University of Washington, Seattle, Washington 98195-7962, USA. hlai@u.washington.edu

In previous research, we found that rats acutely (2 hr) exposed to a 60-Hz sinusoidal magnetic field at intensities of 0.1-0.5 millitesla (mT) showed increases in DNA single- and double-strand breaks in their brain cells. Further research showed that these effects could be blocked by pretreating the rats with the free radical scavengers' melatonin. **Microwave radiation produces DNA breaks in brain cells. This effect could be prevented by using melatonin.**

Environ Health Perspect. 2004 May; 112(6):687-94 **Magnetic-field-induced DNA strand breaks in brain cells of the rat. Lai H**, **Singh NP**.

Bioelectromagnetics Research Laboratory, Department of Bioengineering, University of Washington, Seattle, Washington 98195-7962, USA. hlai@u.washington.edu

Microwave radiation produces DNA breaks in brain cells.

Bioelectromagnetics. 1995; 16(3):207-1 **Acute low-intensity microwave exposure increases DNA single-strand breaks in rat brain cells. , <u>Singh NP</u>.**

Department of Pharmacology, University of Washington, Seattle 98195, USA.

Note: low level microwave exposure results in DNA damage to the brain. DNA damage is associated with an increased incidence of brain cancers.

Arch Med Res. 2005 Jul-Aug;36(4):350-5. **Oxidative damage in the kidney induced by 900-MHz-emitted mobile phone: protection by melatonin.**

<u>Oktem F</u>, <u>Ozguner F</u>, <u>Mollaoglu H</u>, <u>Koyu A</u>, <u>Uz E</u>.
Department of Pediatric Nephrology, School of Medicine, Suleyman Demirel University, Isparta, Turkey. Decrease in renal SOD, CAT, GSH-Px activities demonstrated the role of oxidative mechanism induced by 900-MHz mobile phone exposure, and melatonin, via its free radical scavenging and antioxidant properties, ameliorated oxidative tissue injury in rat kidney. CONCLUSIONS: These results show that melatonin may exhibit a protective effect on mobile phone-induced renal impairment in rats.

Note: Melatonin protects from kidney damage induced by mobile phones.

Mol Cell Biochem. 2005 Aug; 276(1-2):31-7.

Comparative analysis of the protective effects of melatonin and caffeic acid phenethyl ester (CAPE) on mobile phone-induced renal impairment in rat.

<u>Ozguner F</u>, <u>Oktem F</u>, <u>Armagan A</u>, <u>Yilmaz R</u>, <u>Koyu A</u>, <u>Demirel R</u>, <u>Vural H</u>, <u>Uz E</u>.
Department of Physiology, School of Medicine, Suleyman Demirel University, P. K. 13 32100 Isparta, Turkey. drmfehmi@yahoo.com Melatonin and caffeic acid phenethyl ester (CAPE), a component of honeybee propolis, were recently found to be potent free radical scavengers. . Furthermore, treatment of EMR exposed rats with melatonin increased activities of SOD and GSH-Px to higher levels

than those of control rats. In conclusion, melatonin and CAPE prevent renal tubular injury by reducing oxidative and stress protects the kidney from oxidative damage induced by 900 MHz mobile phone. Nevertheless, melatonin seems to be a more potent antioxidant compared with CAPE in kidney. (Mol Cell Biochem 276: 31-37, 2005).

Note: melatonin and caffeic acid were both found to protect animals from microwave induced kidney damage. Melatonin was found to be slightly superior to caffeic acid.

ZINC:

Wei Sheng Yan Jiu. 2000 May 30;29(3):129-31.

[Zinc protective effects on pig retinal pigment epithelial cell damage of lipid peroxide induced by 2450 MHz microwave]

Deng Z, Liu X, Chen J, Yang R.

Faculty of Military Health Service and Statistics, Fourth Military Medical University, Xi'an 710033, China.

Administration of Zn alleviated the increase of MDA and decrease of SOD. In the experiment, 2450 MHz microwave induces the lipid peroxide damage in RPE cells. Zn can enhance the antioxidation ability of cells and alleviate the damage to some extent.

Note: elevated levels of zinc in the blood tend to preserve SOD levels.

S.O.D:

Physiol Genomics. 2003 Dec 16;16(1):29-37

Life-long reduction in MnSOD activity results in increased DNA damage and higher incidence of cancer but does not accelerate aging.

Van Remmen H, Ikeno Y, Hamilton M, Pahlavani M, Wolf N, Thorpe SR, Alderson NL, Baynes JW, Epstein CJ, Huang TT, Nelson J, Strong R, Richardson A.

Department of Cellular and Structural Biology at the University of Texas Health Science Center at San Antonio, San Antonio 78229-3900, USA.

Note: Decreased levels of SOD are associated with an increase incidence of DNA damage and a higher levels of cancer.

Med Pr. 2002;53(4):311-4 **[Effect of electromagnetic field produced by mobile phones on the activity of superoxide dismutase (SOD-1) and the level of malonyldialdehyde (MDA)--in vitro study]**

Stopczyk D, Gnitecki W, Buczynski A, Markuszewski L, Buczynski J.

Zakladu Medycyny Zapobiegawczej i Promocji Zdrowia, Wojskowej Akademii Medycznej w Lodzi. darstop@poczta.onet.pl

The aim of the study was to assess in vitro the effect of electromagnetic field produced by mobile phones on the activity of superoxide dismutase (SOD-1) and the level of malonyldialdehyde (MDA) in human blood platelets. The suspension of blood platelets was exposed to the electromagnetic field with the frequency of 900 MHz for 1, 3, 5, and 7 min. Our studies demonstrated that microwaves produced by mobile phones significantly depleted SOD-1 activity after 1, 5, and 7 min of exposure and increased after 3 min in comparison with the control test. There was a significant increase in the concentration of MDA after 1, 5, and 7 min and decrease after 3 min of exposure as compared with the control test.

Note: direct microwave radiation exposure from cell phones decreased SOD LEVELS while at the same time increased the amount of breakdown products of cell fats. These effects were evident after as little as one minute of phone exposure.

Microwave radiation dropped protective SOD levels as quickly as after only one minute of cell phone exposure.

Head Neck. 1999 Aug;21(5):467-79

Reactive oxygen metabolites, antioxidants and head and neck cancer.

Seidman MD, Quirk WS, Shirwany NA.

Department of Otolaryngology-Head and Neck Surgery, Henry Ford Hospital, 6777 W. Maple Road, W. Bloomfield, MI 48323, USA.

ROM scavengers and blockers have shown both in vivo and in vitro effects of attenuating the toxicity of ROM. Such compounds include the antioxidant vitamins (A, C, and E), nutrient trace elements

(selenium), enzymes (superoxide dismutase, glutathione peroxidase, and catalase), hormones (melatonin), and a host of natural and synthetic compounds (lazaroids, allopurinol, gingko extract).
Note: Antioxidants, catalase, SOD, Gluthatione as well as melatonin and gingko have all been shown to be protective against head and neck cancers.

Antioxidant enzymes in RBCs as a biological index of age related macular degeneration.
Prashar S, Pandav SS, Gupta A, Nath R. Department of Biochemistry, Postgraduate Institute of Medical Education and Research, Chandigarh, India.
The present study was undertaken to assess the levels of antioxidant enzymes in red blood cells of subjects with age-related macular degeneration and age-matched controls. The results obtained show a significant decrease in activities of superoxide dismutase ($p < 0.001$) and glutathione peroxidase ($p < 0.001$) as compared to the controls
Note: individuals with macular degeneration had lower levels of anti-oxidants, SOD or gluthatione as compared to controls, those without the disease but about the same age.

Doc Ophthalmol. 2003 Mar; 106(2):129-36 **Nitric oxide and lipid peroxidation are increased and associated with decreased antioxidant enzyme activities in patients with age-related macular degeneration. Evereklioglu C, Er H, Doganay S, Cekmen M, Turkoz Y, Otlu B, Ozerol E.**
Department of Ophthalmology, Erciyes University Medical Faculty, Kayseri, Turkey. evereklioglu@hotmail.com
BACKGROUND: Nitric oxide (NO), hydroxyl radical (OH*), superoxide anion (O2-) and hydrogen peroxide (H2O2) are free-radicals released in oxidative stress. Superoxide dismutase (SOD), glutathione peroxidase (GSHPx) and catalase (CAT) are antioxidant enzymes, mediating defense against oxidative stress. Excess NO and/or defective antioxidants cause lipid peroxidation, cellular dysfunction and death. Age-related maculopathy (ARM) or

degeneration (ARMD) is the leading cause of irreversible blindness in developed countries.

Note: free radicals contribute to the progression of macular disease and degeneration. Refer section two.

Physiol Genomics. 2003 Dec 16;16(1):29-37 **Life-long reduction in MnSOD activity results in increased DNA damage and higher incidence of cancer but does not accelerate aging.** Van Remmen H, Ikeno Y, Hamilton M, Pahlavani M, Wolf N, Thorpe SR, Alderson NL, Baynes JW, Epstein CJ, Huang TT, Nelson J, Strong R, Richardson A.

Note: Lowered levels of SOD are associated with higher levels of cancer and genetic damage.

GINGKO BILBOA:

Clin Chim Acta. 2004 Feb;340(1-2):153-62
Ginkgo biloba prevents mobile phone-induced oxidative stress in rat brain.
Ilhan A, Gurel A, Armutcu F, Kamisli S, Iraz M, Akyol O, Ozen S.
Department of Neurology, Inonu University, Turgut Ozal Medical Center, 44069 Malatya, Turkey. ailhan@inonu.edu.tr
BACKGROUND: The widespread use of mobile phones (MP) in recent years has raised the research activities in many countries to determine the consequences of exposure to the low-intensity electromagnetic radiation (EMR) of mobile phones. RESULTS: Oxidative damage was evident by the: (i) increase in malondialdehyde (MDA) and nitric oxide (NO) levels in brain tissue, (ii) decrease in brain superoxide dismutase (SOD) and glutathione peroxidase (GSH-Px) activities and (iii) increase in brain xanthine oxidase (XO) and adenosine deaminase (ADA) activities. These alterations were prevented by Gb treatment. Furthermore, Gb prevented the MP-induced cellular injury in brain tissue histopathologically. CONCLUSION: Reactive oxygen species may play a role in the mechanism that has been proposed to explain the biological side

effects of MP, and Gb prevents the MP-induced oxidative stress to preserve antioxidant enzymes activity in brain tissue.

Note: Ginkgo was shown to protect brains from oxidative damage from cell phone exposure and maintained healthy levels of antioxidants.

Jpn J Ophthalmol. 2004 Sep-Oct;48(5):499-502

Effects of oral Ginkgo biloba supplementation on cataract formation and oxidative stress occurring in lenses of rats exposed to total cranium radiotherapy.

Ertekin MV, Kocer I, Karslioglu I, Taysi S, Gepdiremen A, Sezen O, Balci E, Bakan N.

Department of Radiation Oncology, Ataturk University, Faculty of Medicine, 25240 Erzurum, Turkey. mvertekin@hotmail.com.

GB supplementation significantly increased the activities of SOD and GSH-Px enzymes and significantly decreased the MDA level. Total cranium irradiation of 5 Gy in a single dose promoted cataract formation, and GB supplementation protected the lenses from radiation-induced cataracts. CONCLUSIONS: We suggest that Ginkgo biloba is an antioxidant that protects the rat lens from radiation-induced cataracts.

Note: Ginkgo prevented the formation of cataracts induced by radiation.

Fundam Clin Pharmacol. 2003 Aug;17(4):405-17

Ginkgo biloba extracts and cancer: a research area in its infancy.

DeFeudis FV, Papadopoulos V, Drieu K.

Institute for BioScience, 153 West Main Street, Westboro, MA, USA. defeudis@gis.net

Recent studies conducted with various molecular, cellular and whole animal models have revealed that leaf extracts of Ginkgo biloba may have anticancer (chemopreventive) properties that are related to their antioxidant, anti-angiogenic and gene-regulatory actions.

Note: Ginkgo has anticancer, blood vessel protection and offsets DNA damage.

Free Radic Res. 2000 Dec;33(6):831-49
mRNA expression profile of a human cancer cell line in response to Ginkgo biloba extract: induction of antioxidant response and the Golgi system.
Gohil K, Moy RK, Farzin S, Maguire JJ, Packer L.
Department of Molecular and Cell Biology, Lawrence Berkeley National Laboratory University of California, 94720, USA. kishor@socrates.berkeley.edu
Note: Ginkgo biloba tends to protect cells from cancer promotion and increases the anti-oxidant levels in the cell.

BILBERRY EXTRACT:

Adv Gerontol. 2005;16:76-9
[Dietary supplementation with bilberry extract prevents macular degeneration and cataracts in senesce-accelerated OXYS rats]
Fursova AZh, Gesarevich OG, Gonchar AM, Trofimova NA, Kolosova NG.
The testing at 3 month have showed that more then 70% of control OXYS rats had cataract and macular degeneration while the supplementation of BE (bilberry extract)completely prevented impairments in the lenses and retina.
Note: bilberry extract helps to protect and prevent the formation of cataracts and macular degeneration.

CAFFEIC ACID:

Mol Cell Biochem. 2005 Sep;277(1-2):73-80
A novel antioxidant agent caffeic acid phenethyl ester prevents long-term mobile phone exposure-induced renal impairment in rat. Prognostic value of malondialdehyde, N-acetyl-beta-D-glucosaminidase and nitric oxide determination.
Ozguner F, Oktem F, Ayata A, Koyu A, Yilmaz HR.

Department of Physiology, School of Medicine, Suleyman Demirel University, P. K. 13, Isparta, 32100, Turkey. drmfehmi@yahoo.com
In conclusion, the increase in NO and MDA levels of renal tissue, and in urinary NAG with the decrease in renal SOD, CAT, GSH-Px activities demonstrate the role of oxidative mechanisms in 900 MHz mobile phone-induced renal tissue damage, and CAPE, via its free radical scavenging and antioxidant properties, ameliorates oxidative renal damage. These results strongly suggest that CAPE exhibits a protective effect on mobile phone-induced and free radical mediated oxidative renal impairment in rats.

Note: this study shows that caffeic acid which is found in apple juice and bee propolis has a protective effect against the effects of microwave radiation exposure from cell phones. It limited kidney damage in the animals studies and was protective.

CATALASE:

Science. 2005 Jun 24; 308(5730):1909-11. Epub 2005 May 5
Extension of murine life span by overexpression of catalase targeted to mitochondria.
Schriner SE, **Linford NJ**, **Martin GM**, **Treuting P**, **Ogburn CE**, **Emond M**, **Coskun PE**, **Ladiges W**, **Wolf N**, **Van Remmen H**, **Wallace DC**, **Rabinovitch PS**.
Department of Genome Sciences, University of Washington, Seattle, WA 91895, USA.

To determine the role of reactive oxygen species in mammalian longevity, we generated transgenic mice that overexpress human catalase localized to the peroxisome, the nucleus, or mitochondria (MCAT). Median and maximum life spans were maximally increased (averages of 5 months and 5.5 months, respectively) in MCAT animals. Cardiac pathology and cataract development were delayed, oxidative damage was reduced, H_2O_2 production and H_2O_2-induced aconitase inactivation were attenuated, and the development of mitochondrial deletions was reduced.

Note: this study showed that the levels of catalase are critical in preventing disease.

Breast Cancer Res Treat. 2000 Jan; 59(2):163-70   **Lipid peroxidation, free radical production and antioxidant status in breast cancer.**   <u>Ray G</u>, <u>Batra S</u>, <u>Shukla NK</u>, <u>Deo S</u>, <u>Raina V</u>, <u>Ashok S</u>, <u>Husain SA</u>.

Department of Biosciences, Jamia Millia Islamia, New Delhi, India.

Reactive oxygen metabolites (ROMs), including superoxide anion ($O2^*-$), hydrogen peroxide ($H2O2$) and hydroxyl radical ($^*OH$), play an important role in carcinogenesis. There are some primary antioxidants such as superoxide dismutase (SOD), glutathione peroxidase (GPx) and catalase (CAT) which protect against cellular and molecular damage caused by the ROMs. We conducted the present study to determine the rate of $O2^*-$ and $H2O2$ production, and concentration of malondialdehyde (MDA), as an index of lipid peroxidation

Note: this article discusses that increased free radicals increase the rate of cancer production. Also gluthathione, catalase and SOD help to decrease these levels and protect the cell.

FASEB J. 2000 Feb;14(2):312-8   **Oxidative damage to mitochondrial DNA is inversely related to maximum life span in the heart and brain of mammals.**   <u>Barja G</u>, <u>Herrero A</u>.

Department of Animal Biology-II (Animal Physiology), Faculty of Biology, Complutense University, Madrid 28040, Spain.

Note: this study examines the simple fact that increased damage to the mitochondria of the brain and heart accelerate aging in the individual.

CoQ10:

Ophthalmologica. 2005 May-Jun;219(3):154-66

Improvement of visual functions and fundus alterations in early age-related macular degeneration treated with a combination of acetyl-L-carnitine, n-3 fatty acids, and coenzyme Q10.

<u>Feher J</u>, <u>Kovacs B</u>, <u>Kovacs I</u>, <u>Schveoller M</u>, <u>Papale A</u>, <u>Balacco Gabrieli C</u>.

Ophthalmic Neuroscience Program, Department of Ophthalmology, University of Rome 'La Sapienza', Rome, Italy. j.feher@libero.it
The aim of this randomized, double-blind, placebo-controlled clinical trial was to determine the efficacy of a combination of acetyl-L-carnitine, n-3 fatty acids, and coenzyme Q10 (Phototrop) on the visual functions and fundus alterations in early age-related macular degeneration (AMD).

Note: this study demonstrates that maintaining elevated blood levels tends to slow down disease progression. See chapter on antioxidants for a discussion why this is important.

ACETYL-CARNITINE/R-LIPOIC ACID:

Ann N Y Acad Sci. 2002 Apr; 959:133-66
Delaying brain mitochondrial decay and aging with mitochondrial antioxidants and metabolites.
Liu J, **Atamna H**, **Kuratsune H**, **Ames BN**.
Division of Biochemistry and Molecular Biology, University of California, Berkeley, California 94720, USA.
Mitochondria decay with age due to the oxidation of lipids, proteins, RNA, and DNA. Some of this decay can be reversed in aged animals by feeding them the mitochondrial metabolites acetylcarnitine and lipoic acid.

Note: that providing adequate anti-oxidants to protect the mitochondria(cell furnace) is critical in delaying aging.

LYCOPENE:

Clin Chim Acta. 2005 Jul 1; 357(1):34-42
Lycopene but not lutein nor zeaxanthin decreases in serum and lipoproteins in age-related macular degeneration patients.
Cardinault N, **Abalain JH**, **Sairafi B**, **Coudray C**, **Grolier P**, **Rambeau M**, **Carre JL**, **Mazur A**, **Rock E**.
Unite des Maladies Metaboliques et Micronutriments, INRA Clermont-Ferrand/Theix, 63122 St Genes Champanelle, France. Nicolas.Cardinault@clermont.inra.fr
BACKGROUND: Epidemiological studies have established that a low serum concentration of carotenoids was associated with risk of

Age-Related Macular Degeneration (ARMD). Individual carotenoid levels showed that only lycopene was decreased significantly in serum, LDL and HDL fractions in patients ($P<0.05$).

Note: Lycopene appears to be used up. Lycopene is an antioxidant which is found in red vegetables especially tomatoes.

AUTOBIOGRAPHY

I am a medical internist with over 30 years of clinical experience. Additionally I have been researching the biomedical literature from a traditional and holistic perspective for the last 15 years. I have training in herbology, Chinese medicine, nutritional approaches to health and other modalities. I have done extensive research on the causality of disease and its relationship to environmental factors as will as researching approaches to deal with these issues. I have been a radio show host involved in sharing information with the public on alternative health issues and perspectives in the NYC area for the past 5 years and have been actively involved with a nutritional pharmacy.

Printed in the United States
59374LVS00002B/16-63

9 781425 904807